Raptors of the Rockies

Biology of the birds of prey and species accounts of the raptors of the Rockies

Kate Davis

Creative consulting and design by Ken Lockwood

Photos, etchings, and drawings
by Kate Davis
unless otherwise noted.

Raptors of the Rockies

2002
Mountain Press Publishing Company
Missoula, Montana

Cover photos: Rough-legged Hawk, Short-eared Owl, Gyrfalcon,
Great Horned Owl by Kate Davis. Bald Eagle © Milo Burcham

All photographs and illustrations by the author unless otherwise credited.
Range maps © Bob Petty

Library of Congress Cataloging-in-Publication Data

Davis, Kate
  Raptors of the Rockies : biology of the birds of prey and species
accounts of the raptors of the Rockies /Kate Davis.
      p. cm.
Includes bibliographical references and index.
  ISBN 0-87842-461-X (alk. paper)
 1. Birds of prey--Rocky Mountains. I. Title.
  QL677.78 .D38 2002
  598.9'0978--dc21
2002004917

PRINTED IN HONG KONG BY MANTEC PRODUCTION COMPANY

Mountain Press Publishing Company
P.O. Box 2399 • Missoula, Montana 59806
406-728-1900

Dedicated to my Mother, Sally Phillips

Table of Contents

Foreword

Kate Phillips Davis

With the raptors that Kate Davis keeps for education, she has mastered the art of bringing a bit of the wild into the classroom. There is nothing quite like the experience of observing wild animals up close, especially species that one rarely sees except at a distance, if even that! Through Raptors of the Rockies programs, people become transfixed by the presence of Max, the Golden Eagle, and are left speechless by the miniscule size of DotCom, the Northern Pygmy-Owl.

Perhaps because of her formal education in biology (she has a B.A. in zoology), Kate conducts education programs that don't mislead the public into viewing wild animals as pets. She is able to focus her monologue on the biology of raptors, while at the same time using humor to carry her audience along as she informs them about the way each species is wonderfully adapted to live in the wild. The dedication and care Kate gives her birds is apparent, as is the creative effort she brings to naming each of her birds!

I'm delighted to see this book; it's a perfect companion for Kate's public presentations, and it stands well on its own as a guide to raptors of the Rockies. Kate's photography, artwork, and text provide a wonderful stimulus to learn more about this most captivating group of birds.

Dr. Richard L. Hutto
University of Montana
Host of *BirdWatch*, PBS Television

Kate Phillips Davis

Acknowledgements:

We would like to thank the following folks for helping make this book possible :

    Ken Lockwood
    Jeff Marks
    Brigid Wilson
    and Robert Petty for the range maps

Introduction

As the only wild animals that we see all the time, birds are our constant companions outdoors. We see songbirds and woodpeckers enjoying the bird feeder, ducks and geese at the lake, and pigeons and starlings that live downtown. We are lucky to catch a glimpse of other animals, like a coyote or a rattlesnake. More commonly we may spot a deer in a hay field. So birds are our most conspicuous, or easily seen, wild neighbors.

Raptors are birds of prey, meaning that they make their living hunting other animals. Some species may be found in open country or deserts, others in dense woodlands, and still others around waterways. Each species has traits that reveal a unique adaptation for predation.

Are all meat-eating birds raptors? No. For example, herons, kingfishers, and mergansers all eat fish, which is meat. And the insect feeders like warblers and swallows are meat eaters too. These birds are predators in their own way, but they are not raptors.

The term "raptor" is applied only to two groups of birds:

The Order Falconiformes, or hawks, eagles, falcons, and Ospreys—
the daytime, or diurnal predators

and

The Order Strigiformes, the owls—
mostly nighttime, or nocturnal, predators

Taxonomy helps scientists divide these birds into two large groups.

This is an example of how a bird may be classified:

Kingdom - Animalia, the animals

Phylum - Chordata, animals with a notochord associated with backbones

Class - Aves, the birds

Order - Falconiformes, the hawks, eagles, falcons, and Ospreys

Family - Falconidae, the falcons

Genus - *Falco*

Species - *sparverius*

the American Kestrel

The scientific name of an animal is composed of a genus and species, the words often being Latin or Greek and meaning something specific. *Falco* is falcon and *sparverius* means "pertaining to a sparrow." These birds were formerly called Sparrow Hawks.

All living beings—plants and animals—may be organized into groups like these so that we may better understand how they are related to each other.

Raptors as Hunters

Harris's Hawk

E ven though hawks and owls are not closely related, they have many structures and behaviors in common. This is called convergent evolution because both groups have evolved similar traits that enable them to hunt other animals.

Find:
All raptors must first locate their prey, and their senses are very well developed for this.

Gyrfalcon

Vision:
Raptors have binocular vision like people, which means that their eyes face forward and have an overlapping view that gives them good depth perception. That isn't to say that they are seeing through binoculars, as they can see close-up quite well. On the back surface inside the eye the picture is formed on the retina. Raptors have about five times more photoreceptive cells on the retina than humans. That's like having more pixels on a computer picture. The image a raptor sees is that much clearer. A Golden Eagle can see a rabbit that is two miles away. Owls can see in low light with their huge eyes and wide pupils, and with a greater number of special structures called rods. All birds have a third eyelid, called a nictitating membrane. This is very obvious in owls when they blink their upper eyelid and nictitating membrane. In the hawks it is a flash of blue-gray as they blink. This membrane keeps the eye from drying out and protects the surface, as when raptors strike their prey.

Hearing:
Owls have excellent hearing. For the nocturnal species, their well-developed ears allow them to hunt in near darkness. The facial disks that surround their eyes first direct the sound to the huge ear openings. These openings are asymmetrical, with one placed higher and one lower on the head. This allows them to lock in on the source of the sound by triangulation. Hearing is less important to the hunting hawks, but they can recognize individuals and offspring by voice alone. Cooper's Hawks listen for the calls of quail. And harriers have facial disks like owls and hunt using their hearing more than most other hawks.

Smell:
Some North and South American vultures (genus *Cathartes*) and some sea birds may be the only birds in the world that have a well-developed sense of smell. This may help them locate their food.

Bird Brains:
All birds have keen, quick brains that allow them to live in the three-dimensional world of flying left and right, up and down. The human brain can comprehend 20 images per second, so films and cartoons play 24 pictures per second and our brains perceive that as motion. Birds can understand 70 to 80 images per second, so they are living in a much faster world than people. In this way they can fly though the trees at high speed or take off in a flock without bouncing off each other. A bird brain may be about ten times faster than that of a human. Raptors, like other visual animals, may have a search image in their brains, or picture of the prey, and know what they are seeking in their hunts. Plus they can detect motion very quickly for fast reactions to the quarry.

Northern Saw-whet Owl

Catch:

The techniques of catching other animals come in two types.

Sit and Wait:

This technique is the most energy efficient, where the raptor may find a vantage point like a tree or power pole and wait for the food to show up. The producers, or prey, may simply appear on the hunting grounds of the consumers, or predators, at which time a quick chase may result in capture or escape. In most cases the prey is common and mobile, and sitting quietly at a good observation point often pays off for the predator. Accipiters like Sharp-shinned and Cooper's Hawks will wait for a group of seed-eating birds to show up at a bird feeder, then dash out to grab one out of the air. Kestrels drop to the ground from a phone line for grasshoppers and mice. Owls may pounce from a perch at night.

Rough-legged Hawk

Active Search:

This other hunting strategy is also widely used. It may take a lot of energy, but could mean a great meal. Golden Eagles may fly low over hunting grounds and falcons may chase a bird in the air, or drop down and hit it in a stoop. The chase is swift and they know when to give up. Success for the predator is more difficult because the intended prey would like to continue living and doesn't want to get caught. Target animals have a wide range of tactics for getting away. Ground squirrels will dive down a burrow, rabbits have incredible stamina during a chase, fish swim for deeper water, and zigzagging birds in flight make a tough target. The safety-in-numbers theory can be seen when a loose flock of birds tightens up with the appearance of a falcon, making prey selection more difficult. A greater number of alert eyes in the flock or colony makes a sneak attack harder to pull off. The search-and-chase tactic is often mixed with the sit and wait for good results for the successful predator.

Northern Boreal Forest

Kill:

Although many individuals eat carrion—animals found dead—all North American raptors are experts at killing.

Structures:

The primary weapons of the raptors are their feet, specifically their talons, or curved toenails. The back toe, or hallux, pierces the prey in hawks, and owls and Ospreys can rotate their outer toe to hit with two toes in front and two in back. These powerful feet grasp the prey while they tear with sharp, hooked beaks, which are often killing devices. Some owls will crush their prey with their feet, but many raptors deliver the death blow with a bite to the back of the neck. They may swallow their prey whole, as with small rodents, or tear off bite-sized pieces of meat using this specialized beak. They often mantle their meal, covering it with outstretched wings to hide it from other predators that may want to steal it.

Predation:

Raptors are predators, like their mammal counterparts the carnivores. Life is very challenging for the raptors, and relatively few make it to breeding age. Many things can go wrong in the process of finding, catching, and killing prey. They can hit windows, cars, and power lines in populated areas, or other objects out in the wild. The prey item may fight back, causing injury, or the raptors themselves may fall prey to another predator while roosting or hunting. Weather such as blizzards and high winds can play a role, and weakened birds may die of disease. And unfortunately many are killed outright by people through gunshots. Mortality, or death rates, for hawks and owls may be as high as 60-70% over the first year. That would mean that of 100 baby hawks or owls hatched in the spring perhaps only 35 would be alive a year later. Mortality rates drop as the birds get older and gain experience. These birds have learned how to hunt and migrate, where to roost, and how to be successful as a raptor. The larger species like eagles and Great Horned Owls live longer and have a lower death rate as adults. They can produce young over a longer period of time and have a smaller number of eggs per nest. With relatively less energy spent in reproduction, they can invest more energy on their own well being, and so have a longer life-span. Smaller species don't live as long, and produce more eggs for each nest. They may produce a great number of offspring in a short time, but this takes more energy. So these birds are short-lived.

Success rates for catching different prey types vary. Insects may be easy to catch and mammals and fish harder to catch. Probably the hardest to kill are birds in the air. Perhaps only 7% of these bird chases result in a meal. Those animals that are eaten are often the ones that are most easily caught, so are usually individuals that are sick, weak, malformed, or the young and very old. There are always more prey animals than predators, so the raptors can keep these numbers in check. For example, if a pair of robins had two broods a year and none of the offspring died, those would breed and in ten years would result in 19 million robins. Predation is an important way to keep these numbers in check.

Five-week-old Red-tailed Hawk

Raptors of the Rockies

Successful predators should minimize aggressive encounters, or interference competition with the other animals that want to eat the same things they do. One way to do this is by being active at different times of the day. Daytime raptors are diurnal versus the large number of nocturnal owls that hunt at night. Although these owls can see very well during the day, the hawks can't see at night so owls have a distinct advantage. These alternate periods of activity make sure that they don't compete aggressively with each other. Crepuscular animals are active at twilight—dawn and dusk. Also, the habitats that raptors live in may be exclusive in that forest hawks will eat different animals than do open-country soaring hawks like Red-tails. Ospreys eat just live fish, but Great Horned Owls will eat just about anything, so are generalists.

Raptors have various food requirements, depending upon size. The largest species such as Golden Eagles may eat only 6% of their body weight daily, and after a big gorge, may not feed for several days. The smallest owls, on the other hand, will eat half their body weight daily. This has to do with metabolism, surface area, and body heat loss. The smaller the raptor, the more food it eats as a percentage of body weight. The weather also plays an important part in diet. Colder temperatures mean that it takes more energy to stay warm, meaning more food. Raptors may eat 20% more during cold months and have the same body weight. Maintaining the 102 degree plus Fahrenheit body temperature of a bird takes a lot of energy. Also, owls don't have a crop, or pouch, in their throat in which to store food, unlike other birds. Hawks can fill that crop and digest slowly at their leisure. Owls often cache, or hide, the uneaten portions of an animal and return later to eat what's left. This is also a common trait of falcons, but probably all the birds of prey will cache food at some time.

Another rather unique trait of raptors related to predation is called reversed sexual dimorphism. This means that the females are larger than the males in most species. In the falcons for example, the males are called tiercels, meaning one-third smaller. The accipiter females may be twice the weight of the males. This size difference is especially great in raptors that eat birds. There are many theories as to why this is true. The most accepted is that the female does all of the nest-work early on, the incubating and feeding of tiny young. The male is smaller and brings smaller prey items, like songbirds. He may be more agile and maneuverable in the air due to his size. The female will finally leave the nest and hunt when the growing young demand more food than the male can supply. So she may travel longer distances for larger prey. It may mean the difference between a small songbird and a large grouse to feed the family. Again, this is only a guess as to why these females are larger. Those species that feed on insects, such as small owls, are the same size, and with vultures the males may be larger.

Asychronous hatching of Northern Harrier chicks

Raptors of the Rockies

Migration:

Migration is important to many raptor species. In North America it is usually a movement south in the fall, and back to the breeding grounds north in the spring. Fall movements are more concentrated and are quite often associated with the onset or passage of a cold front. Large numbers of raptors may be seen on certain migration paths. These may be favorable because the contour of the land creates wind currents, such as the Rocky Mountain front in Montana. These winds and thermals make flying long distances easier. Swainson's Hawks hold the migration record for North American raptors, as they leave the U.S. and spend the winters in South America, 7,000 miles distant. The entire population of Rough-legged Hawks leaves their arctic breeding grounds to over-winter in the United States and Canada. Spring migration back north is more dispersed, perhaps because there are more thermals over a larger area. Other seasonal movements may be local, or of a relatively short distance to better hunting grounds after the breeding season, like the Prairie Falcon's. Periodic irruptions occur when prey numbers drop dramatically and large numbers of raptors move further distances than they normally would, even in all directions. Vole numbers crash about every 3-5 years, and snowshoe hares every 7-10 years, so we see more Northern Goshawks and Snowy Owls than usual during these irruptive years. Still other birds are nomadic, moving in all directions seeking good hunting areas and prey. Short-eared Owls and Rough-legged Hawks are sometimes nomadic.

Rough-legged Hawk

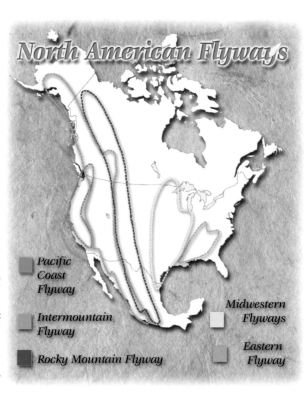

North American Flyways

- Pacific Coast Flyway
- Intermountain Flyway
- Rocky Mountain Flyway
- Midwestern Flyways
- Eastern Flyway

This Golden Eagle was banded and tagged with a patagial marker to monitor its movement from Rogers Pass in Montana. The colored markers are numbered and are relatively easy to read, plus they apparently don't interfere with flight.

Young Northern Goshawk

The name *Raptor* comes from ancient Latin and means "to plunder" or "to seize and carry away". Indeed these kinds of birds do both. The raptors rely on the meat of other animals to live.

The Falconiformes, or daytime raptors, can be classified, or grouped like this:

Order Falconiformes:

Hawks, Eagles, Falcons, and Ospreys
Family Accipitridae:

Genus   ***Buteos***
Red-tailed Hawk
Rough-legged Hawk
Swainson's Hawk
Ferruginous Hawk

Red-tailed Hawk

These are the soaring hawks that take advantage of the thermals, or the wind currents, that occur when the ground air has heated up and is lighter, causing updrafts. Buteos often hold their wings in a dihedral, or little V when soaring. They can spot prey from a long distance and also still-hunt from perches such as power poles and tree snags. The Buteos are large-bodied birds with broad wings and tails, and have what we call low wing loading. This means that they have a large surface area of wings and tail compared to their weight. The buteos are mostly mammal eaters, feeding on rodents and rabbits plus some birds and reptiles.

Genus *Aquila*, the Booted Eagles
Golden Eagle

Golden Eagle

Golden Eagles are large open-country raptors that feed mostly on squirrels, rabbits, and medium and large birds. Golden Eagles build bulky stick nests on cliffs and occassionally in large trees, and may fledge one or two young. They may hunt from a high soar, or low in flights that hug the contour of the surface, surprising their quarry as they suddenly appear over a hill. Many birds rely on carrion in the winter. Goldens range in size from between 8 pounds for males to 12 pounds for females in America, and the species lives all across the northern hemisphere.

Genus *Haliaeetus,* the **Sea Eagles**,
Bald Eagle

The white head and tail of adult bald eagles makes these huge raptors widely recognized. The subadult forms are not so easy to identify and have a complicated pattern of molts for five or six years. Largely fish-eaters, they are usually seen around rivers and lakes. Bald Eagles also eat carrion, or dead animals, so may be away from water, especially when migrating and in winter. They are also very good at hunting waterfowl. Bald and Golden eagles are about the same size, but Balds hold their wings flatter, or more horizontally, in flight.

Bald Eagle

Cooper's Hawk

Genus *Accipiters*
Sharp-shinned Hawk
Cooper's Hawk
Goshawk

These are the forest hunters that feed largely on birds caught in the air or chased to the ground and pursued through brush. With their short wings and long tails for steering, these raptors may dart through the forest foliage and are usually secretive except in breeding season. The males will do display flights over the terrritory, and both sexes may defend the nest site aggressively. Sharpies may take advantage of a bird feeder as a buffet from which to choose a meal. Evidence of an accipiter may be a plucking post, or favorite spot to dismantle the bird it has killed, with feathers and bones lying about. Mammals like red squirrels are also taken, and Goshawks catch snowshoe hares. They build nests of sticks high in trees, which are often difficult to locate.

Genus *Circus,* the **Harriers**
Northern Harrier

The harriers were formerly called "marsh hawks" and live in the open country of grasslands and meadows. Their prey are small mammals, birds, and reptiles. Like owls, Harriers have facial disks or round ruffs of feathers that circle the face. They fly low over fields and listen and look for their prey with their wings held in a little V dihedral. Harriers have long wings and tails. The females and young birds are a rich brown color, while the males are silver gray with black wing edges. Both sexes have a distinct white rump that is easily seen in flight. The female with her brown coloration blends into the grasses as they nest on the ground. Sometimes one male may have more than one mate, and feed several females and young in the breeding season.

Northern Harrier

Family Pandionidae:
Genus *Pandion,* the Ospreys

One species lives all over the world, except Antarctica. This large raptor feeds only on live fish and has excellent adaptations for catching this slippery prey. They have spicules, or horny spikes on their feet to help hold on to fish, and they may rotate an outer toe so that there are two toes are in front and two behind to better catch and balance this prey. They also have a nostril that can close when they hit the water. Ospreys nest in snags and recently on the tops of power towers and electric poles, and few predators will risk a few million volts to climb to an Osprey nest. Most Ospreys migrate far south to Central and South America in the late fall, and the youngsters remain there for several years, returning to where they were hatched when they are old enough to breed.

Osprey

Peregrine Falcon

Family Falconidae:
Genus *Falco,* the Falcons
American Kestrel
Merlin
Prairie Falcon
Peregrine Falcon
Gyrfalcon

Falcons have long, pointed wings and a long tail, and have high wing loading, or are relatively heavy compared to a smaller surface area. These are the fastest fliers on earth. Peregrine Falcons may stoop at speeds of over 100 miles an hour, hitting birds in the air with closed feet. Merlins and Gyrfalcons chase bird prey in flight with great speed. Prairie Falcons eat ground squirrels in summer, then may catch birds in the air over the winter. The American Kestrel is our most common diurnal, or daytime, raptor, seen all summer perched out in the open on power lines and poles. Falcons have a tomial tooth, or small notch, in their beak for breaking the neck of their prey and tearing apart the tough parts. Another unique structure is a round nostril with a central bony baffle to help falcons breathe during high-speed flight. The skin around their eyes is bare of feathers, and for most species changes color with age. Falcons don't build their own nests. Kestrels use woodpecker holes, Merlins often use magpie nests, and the larger falcons simply have a little scrape on a rocky ledge.

Summer Osprey breeding habitat

Turkey Vulture

Family Cathartidae:
the New World Vultures
Genus *Cathartes,* the Turkey Vulture

Due to studies of DNA, or the genetic makeup, scientists now know that New World Vultures are actually more closely related to storks than hawks. American vultures are not really raptors at all, and have been classified in the stork order since 1998. We include them here because of their historic ties with the hawk order and reference in most field guides. Our only representative of this group in the northwest is the Turkey Vulture, which is a resident only during the summer months. They do a great cleanup job on dead animals. In flight Turkey Vultures may be called TV's, largely because their wings look like a little V from head-on. They have strong beaks for tearing meat, but weak, unraptor-like feet. The *Cathartes* vultures are among the few birds in the world with a well-developed sense of smell, and they can find their food of rotten meat with their nostrils. With food frozen and no thermals, or wind updrafts, for soaring in the winter, Turkey Vultures must migrate south in the fall.

Raptor Habits

Gyrfalcon

Northern Harrier

Prairie Falcon

A Prairie Falcon mantles a freshly caught ground squirrel, hiding it from other raptors that may want to steal it.

Preening is an important daily activity of all birds, and the feathers are kept straight, clean, and oiled from the uropygial gland.

Raptors of the Rockies

Falconiformes Silhouettes

harriers

- Northern Harrier

accipiters

- Sharp-shinned Hawk
- Cooper's Hawk
- Northern Goshawk

buteos

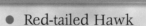

- Red-tailed Hawk
- Swainson's Hawk
- Rough-legged Hawk
- Ferruginous Hawk

eagles

- Golden Eagle
- Bald Eagle

falcons

- American Kestrel
- Merlin
- Prairie Falcon
- Peregrine Falcon
- Gyrfalcon

ospreys

- Osprey

Kate Davis

Order Strigiformes:

Owls

Family Tytonidae:

Barn Owl

Family Strigidae:

Small-eared owls
Great Horned Owl
Northern Pygmy-Owl
Flammulated Owl
Western Screech-Owl
Northern Hawk Owl
Snowy Owl
Burrowing Owl

Large-eared, mostly forest owls
Great Gray Owl
Barred Owl
Northern Saw-whet Owl
Boreal Owl
Long-eared Owl
Short-eared Owl

Alan Nelson

Barn Owl

Great Horned Owl

Western Screech-Owl

Long-eared Owl

Barred Owl

The owls are unique in so many ways:

Nocturnal: Most owls are nocturnal, with physical and behavioral traits that allow for a nighttime lifestyle. Many owls are crepuscular, or more active at dawn and dusk, and others are diurnal, or active during the day. Occasionally, nocturnal owls are caught out in the open during the day and are mobbed by songbirds. The little birds will circle the owl with alarm calls to harass the owl and let others know that a predator is in the area. Small birds will bravely attack a large owl, knowing that they they are too swift and agile in the air for a big predator to catch. This can be a great way to observe an owl during the day, by listening and looking for mobbing songbirds or crows.

Feathers: Most owls have dense, soft, loose feathers. The leading edges of their wings have fringes that help make their flight nearly silent by breaking up the flow of air. In this way they can continue to hear their prey while they fly, and they can't be heard by their intended prey. Some daytime birds, like pygmy-owls, don't really need to be silent, so are as noisy as hawks in flight. Most owls must be inconspicuous and hide while they roost, or spend their non-hunting hours. For this reason they have cryptic coloration or camouflage so that they blend in with their surroundings and don't stand out for the daytime predators to see. The feathers of screech-owls resemble the bark of trees, and the Short-eared Owl's look like grass. The plumage or feather patterns are nearly the same between males and females for all in North America but the Snowy Owl.

Eyes: The eyes of owls are huge and are placed far apart on the front of their head on a relatively flattened skull. A male Great Horned Owl that weighs about two and a half pounds has the same size eyes as an adult human. With forward-facing eyes, owls have a big overlap of left and right eyes to allow them to see in binocular vision. Owls have a 110 degree field of vision (people have 180 degrees), with over half of this as an overlap. This provides the depth perception needed to hunt. Most other birds have eyes on the sides of their head to detect predators, such as raptors. As a way to make their eyes larger and longer to be more effective in the dark, owls have eye bones arranged in tubes that are attached to their skull. These are called sclera. They can't move their eyes in the sockets, but can rotate their head more than 270 degrees (360 is a circle). So for even little changes in view, they must turn their heads. The convex bulge or bubble on the front of the eye or lens allows for a greater surface area to collect light. The pupil can open to nearly fill the iris to collect more light. Owls bob their heads up and down, and back and forth, to view objects from lots of different angles to better judge distance. This is called motion parallax — things that are closer appear to move more than those farther away. Also owls see mostly in shades of black and white instead of brilliant color. That is because of the photo receptors in the eyes. Owls have a great number of rods that are sensitive in low light. People have cones to see in color. Owls can still see very well during the day, better than people, because of the high density of cells in the retina where the picture is formed. But no other avian predators can get around at night like owls. So these species have an advantage by having little competition.

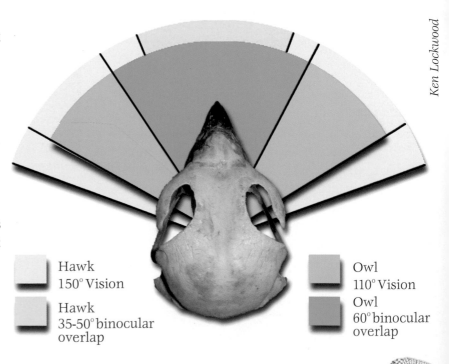

Ken Lockwood

Hawk
150° Vision

Hawk
35-50° binocular
overlap

Owl
110° Vision

Owl
60° binocular
overlap

Ears:
The ear tufts, or groups of feathers on top of the head on many owls, are not really ears; they may be a way to break up the outline of the owl for camouflage. These are only present in forest owls. The true ears are large openings on the sides of the facial disks of feathers. The facial disk is like a satellite dish that gathers sound. Soft feathers in the front allow sound to pass through, and stiff feathers behind trap the sound and direct it to the large ear openings. Owls are unique in that in many species the ear openings are asymmetrical, or uneven. The left opening is lower and the right is higher on the head, as compared to mammal ears which are even on both sides. For an owl, a sound may enter the left side one 30,000th of a second before the right side, and it may hold its head until the sounds are equal. They can triangulate, or zoom in exactly where the sound is on a horizontal and vertical plane. Some owls can hunt in darkness using their hearing alone, especially when the prey is scooting around in the leaves and grass.

Long-eared Owl

Digestion:
Birds have a fast digestion that is necessary for weight reduction and flight. Food passes quickly through their system. Owls are different from other birds in that they don't have a crop or pouch in their throat to store food waiting to go to the stomach. Hawks can gorge on a huge meal and digest the food stored in the crop all day. Owls may swallow their meal of a rodent whole and head-first. Like all birds, they have two stomachs. The gizzard, or second stomach, sorts out all of the digestible food, then packs all of the bones and fur of the prey into a little ball called a pellet. Owls can't digest bones, so about 46% of a pellet is the entire mouse skeleton, for example. The Falconiformes have stronger stomach acids, but they still form pellets or castings of fur or feathers with 6% bones. Pellets and castings scour out raptors' systems and are essential to their health. When they get hungry for another meal, they cough these up, a voluntary regurgitation. This is usually first thing in the morning for hawks and evening for owls. After a few shakes of the head, out of their mouth comes the pellet, and it's time to hunt again. To scientists, a pellet is a little packet of information that tells them just what the raptor has eaten and where it roosts. Much information about owls has been gathered by dissecting these pellets.

Great Horned Owl pellets

Feet:
All owls, like most birds, have four toes. The outer one on each foot can rotate forward, so usually owls have two toes facing forward and two backwards to better grab their prey. This forms a deadly square of talons on each foot. Most owls have feathers all the way down to these talons. Since they usually swallow their prey whole, feeding in this way is very tidy and their feet remain clean. This may be one reason they have feathered tarsi, other than the added comfort of warmth. Tropical owls in warm climates often have bare feet with no feathers.

*Owl's powerful feet
and sharp talons*

Vocalization:

With owls' largely nocturnal habits, calls are more effective than a visual flight display to mark territories. Owls call to attract a mate, maintain a bond between pair-members, or to defend a territory. These calls are distinctive enough that the species can be identified by voice alone. Other sounds they make may be alarm calls to warn of a potential predator, maintain contact with family members, or for youngsters and breeding females to beg for food. Another sound common to all owls is the bill-snap. Owls snap their top and bottom beaks together to make a cracking noise as a warning. Even owls as young as a few days old make this noise. People can imitate distinctive calls and attract wild owls to perch within a few yards, especially in the late winter breeding season of Great Horned and Barred Owls.

Milo Burcham

Nesting:

Owls do not build nests, and more than half of the species are cavity nesters. Usually a nest of another bird is used, and woodpecker holes are highly valued. Flicker cavities are ideal spots for many species. Larger owls use the stick nests of other birds like hawks and crows, or hollows in broken dead trees. They may also nest on cliffs and rock ledges safe from ground predators. Bonding between a pair of owls is strengthened in many species by a process called allopreening. The male and female preen each other, keeping their feathers neat and tidy with their beaks. They also may call or hoot together in a duet, with the male typically starting a song and the female joining in, back and forth, to further encourage a successful pairing.

Great Gray Owl feeds her nestlings.

Great Horned Owl still in downy feathers.

Young:

The small, downy young hatch with their eyes closed and are fed tidbits by the female. As they grow, whole prey items are brought to the nest, which they swallow whole or tear apart. For many owl species, the female begins incubating after the first or second egg is laid. She will continue to lay eggs and incubate as these first chicks are growing. Most owls show asynchronous hatching, with one chick hatching every two days or so. For this reason the young are different ages and sizes. One Short-eared Owl chick, for example, may be fledging, or leaving the nest, just as a nest mate is hatching. They are often fed by the parents, sometimes just the male, for many weeks after fledging. Then they are on their own and may be driven from the territory by their parents, as with Great Horned Owls. Larger and nonmigrating owls may form pairs that last for a lifetime.

Raptors of the Rockies

Migration:

Most North American owls are resident birds year-round. For example, generalist feeders like Great Horned Owls will eat anything and don't rely upon prey that disappears over the winter, like insects. The most migratory is the tiny Flammulated Owl that relies upon insects and migrates to Mexico and Central America. Other birds, like the Northern Pygmy-Owl, move from the mountains into valleys in the fall where the climate is less harsh. And Short-eared Owls may search for prey in all directions and are nomadic.

Cryptic coloration of Flammulated Owl

Convergent Evolution:

A process called convergent evolution has taken place between the two orders of raptors, the Falconiformes and Strigiformes. Both hunt other animals, so they have evolved the same physical traits even though they are not related. Large, forward-facing eyes, decurved sharp beaks, and powerful feet with sharp talons are some of the traits shared between hawks and owls. With a largely nocturnal group, the owls, some counterparts are missing. For example, the large soaring buteo hawks often rely on day-time ground heating to cause thermals for rising wind currents. This doesn't occur at night, so no owls hunt by soaring. Falcons chase down birds in high-speed pursuits. Most prey birds are inactive at night and these chases wouldn't work in the dark. So owls usually have the "sit and wait" tactic of hunting. Perches both in the open and in the forest are good places from which to spot prey. Some open-country owls, like the Short-eared and Snowy, use the "active search" technique, flying back and forth low over fields.

Western Screech-Owl demonstrates head first, swallow-them-whole technique.

Woodland edge habitat of many owl species

Raptors and Humans:

Birds of prey have always held a fascination for people, even before recorded history. Their images appeared in cave paintings, and stories of raptors with magical powers have survived in cultures around the world. Ancient Egyptians had a falcon god and Greeks and Romans worshipped Athena, goddess of wisdom, whose sacred bird is the owl. Birds of prey adorn coins, coats of arms, tapestries, sculptures, paintings and pottery around the world. They have been considered deities or gods, as well as underworld demons. Raptors have been symbols of strength, power, freedom, fidelity, and even war. The Bald Eagle is our national bird. Mexico's national bird is the Golden Eagle, and the Gyrfalcon for Iceland. Birds of prey inspire some feeling from nearly everyone.

Falconry:

The oldest sport in the world is falconry, with a 4000-year-old history. Falconry means "hunting wild game in the field with a captive raptor". It probably began in China when people trapping birds for food realized that raptors also capture game and could be trained. Records from 600 B.C. describe hunting with hawks in India, Japan, and the Middle East. The sport reached its high point in the Middle Ages in Europe when it was the number one sport with a huge following. Social rank or caste dictated what species of bird a person kept. Gyrfalcons were for kings, Peregrines for the Prince, landowners and bishops kept Goshawks, priests had sparrow hawks, and for the lowly knave, a kestrel. Women of the court flew Merlins which chased skylarks higher and higher in "ringing flight". The invention of firearms spelled death to birds of prey, as now they were thought to compete with hunters for game animals. Extermination of all raptors as vermin was the new battle cry. Extensive agriculture and hunting preserves for the very

Gary Lackie

AREA CLOSED

REVEGETATION PROJECT

U.S. Department of the Interior, National Park Service, Denali National Park

✦ U.S. GPO: 1993-791-075

Gyrfalcon

rich created bounties on raptors, or money paid for each one killed. This persecution persisted until very recent times. In 1918 the Migratory Bird Treaty Act protected raptors and other nongame species. The Bald Eagle Protection Act of 1940 was extended to include Golden Eagles in 1962.

Falconry is still very popular in the countries of the Middle East, such as Saudi Arabia. In the U.S. it has a following of dedicated falconers. Strict laws and regulations dictate the sport, and it is only for the most dedicated individuals. Regulations state that only young birds with little hunting experience may be taken from the wild, and are often released after some seasons of hunting. There is no impact on wild populations, and wild Peregrines are still protected. Captive-bred birds are more common now, purchased from breeders. The sport survives, little changed after thousands of years.

Ron Austing

Kate Davis trains a young Harris's Hawk for falconry.

Raptors as Barometers:

Raptors are at the top of the food chain. They are considered "environmental barometers," or the animals that tell us how the natural world is doing with all of the impacts from humankind. What this means is that if raptor populations drop, something is wrong in the environment. Tiny animals like invertebrates are eaten by insects, which are eaten by minnows then by larger and larger fish, then an Osprey. Or insects by birds which are eaten by an accipiter or falcon. The food chain often ends with the raptors. If any poisons are put into the environment, like insecticides to kill the pests that harm agriculture, the poisons stay in the body of each animal, becoming more concentrated with each host. This is called biomagnification so that raptors that eat animals in a long food chain ingest highly concentrated toxins or poisons, which is very unhealthy. Peregrine Falcons caused a huge alarm call about the use of DDT, a pesticide used to kill insects starting in the 1940s. The native subspecies became extinct in the eastern U.S. and numbers dropped in the West. The reason was the spraying of fields with DDT. Peregrines ate birds that had high concentrations of toxins, and it caused their eggshells to be so thin that they couldn't incubate the eggs and the young died. They couldn't reproduce in any healthy numbers. The use of DDT was stopped in 1972 in the United States, and it opened the eyes of the world to

Peregrine Falcon

environmental dangers. So these top predators tell us how the natural world is doing. Barometers are used to forecast the weather, and raptors give a forecast of the environment.

The Peregrine population has made a dramatic comeback through captive breeding, one of the happy endings thanks to concerned people and admirers of this species. The Peregrine Fund has released over 6000 birds hatched in man-made breeding chambers and sent to "hack sites" to naturally reintroduce youngsters to the wild with little human contact. Today the numbers have mostly been restored, and the continental subspecies *Falco peregrinus anatum* was taken off the Endangered Species list in 1999. These have been gigantic efforts to save a spectacular species that truly deserves respect.

Legal Protection:

The Migratory Bird Treaty Act protects all native birds, including raptors. It is against the law to possess native birds, dead or alive. This even includes feathers and nests. The law does allow for the hunting of game birds, like ducks and grouse, with restrictions, including seasons and daily limits. Those not protected are the introduced and non-native birds Pigeons, House Sparrows, and European Starlings. Eagles are further protected by a total of three federal laws. Native American people can have their feathers and parts for religious purposes; otherwise possession is a serious offense.

Special state and federal permits allow for birds to be cared for by rehabilitators, or to be kept in captivity for education. Only qualified individuals with the proper facilities may engage in these activities. Bird parts, like feathers and skins, may only be used for educational purposes under a special permit. Native birds are indeed protected, but only fairly recently.

Hawk Watching:

Large numbers of people participate in hawk watching in the spring and fall at favorable migration spots. The overhead spectacle of species and numbers is addictive, and this sport is gaining in popularity. Certain flyways have been known since the 1800s, such as Kittatinny Ridge in Pennsylvania and New Jersey. This was a shooting lane for tens of thousands of "vermin hawks" and thanks to bird humanitarians was named the world's first Raptor Refuge as Hawk Mountain Sanctuary in 1934. New migration routes are being discovered, especially in the West, and professional ornithologists and amateur birdwatchers alike may delight in this activity. Daily report forms are divided into hour-long periods and the number of raptors by species is recorded. Also noted is wind speed and direction and important weather information, plus number of observers. Through consistent hawk watches, population trends of various raptor species can be observed.

Scientific research benefits greatly from raptor trapping and banding. With nets and bow traps birds are caught and various measurements are recorded. When possible, the age and sex is noted, and a light aluminum band with a number is permanently placed on one leg. The bird is then released and the band may be recovered if the bird is killed or caught again. Only people licensed by the federal government may band birds. Scientists hope to gain a better understanding of the mysteries of migration and the life histories of birds through banding.

Rob Domenech catches an adult Goshawk in a bow net baited with a live pigeon.

Morphological measurements are taken, sex and age is recorded, and the bird is banded.

Rob Domenech

Dave Bittner of the Wildlife Research Institute releases the bird to continue her journey south.

Sharon Fuller releases a banded Red-tailed Hawk.

Rob Domenech

If you recover a banded bird, note the exact location and date, and report band number (recoveries only) to:

BIRD BAND
Laurel, MD 20708
or bandreports@patuxent.usgs.gov
or 1-800-327-BAND (2263)

for more information:
USGS Patuxent Wildlife Research Center
Bird Banding Laboratory
12100 Beech Forest Road
Laurel, MD 20708-4037
(301) 497-5790

Order Falconiformes

Order Falconiformes
(Hawks, Eagles, Falcons, and Osprey)

- **Family Accipitridae** (Hawks and Eagles)
  - *Harriers*
    - Northern Harrier
  - *Accipiters*
    - Sharp-shinned Hawk
    - Cooper's Hawk
    - Northern Goshawk
  - *Buteos*
    - Red-tailed Hawk
    - Swainson's Hawk
    - Rough-legged Hawk
    - Ferruginous Hawk
  - *Booted Eagles*
    - Golden Eagle
  - *Fish Eagles*
    - Bald Eagle

- **Family Falconidae** (Falcons)
  - American Kestrel
  - Merlin
  - Prairie Falcon
  - Peregrine Falcon
  - Gyrfalcon

- **Family Pandionidae** (Ospreys)
  - Osprey

- **Family Cathartidae** (New World Vultures)
  - Turkey Vulture

Latin Name: **Circus cyaneus**

Field Marks: **Length 17–38 inches**
Wing span 38–48 inches

Summer
Year-round
Winter

A slim, medium-sized hawk with long, narrow wings and a long tail. Males gray above and white below with black-tipped wings. Females brown above and white with brown streaks below. White rump, or feathers covering the base of the tail. Juveniles resemble adult females, with buff colored underparts. Spend a great deal of time on the wing, low to the ground, slowly flapping or gliding with wings held in a dihedral, or little V. Perches on ground and fence posts, rarely in trees.

Habitat:
Open country grasslands, fields, meadows and marshes. Used to be called Marsh Hawk, but harrier is the correct term for this group of birds that lives all around the world. The Northern Harrier lives all across the Northern Hemisphere and is called a Hen Harrier in Europe.

Behavior:
Eats wide range of mammals like mice and voles, small to medium-sized birds, and lizards. Has facial disk like owls and may hunt using hearing as well as sight. In winter, may roost communally, or spend the night in groups of several to over one hundred birds.

Nest and eggs:
Ground nest of sticks and grasses built mostly by the female. Male transfers food to female in flight. Aerial prey transfer occurs during incubating and early brood rearing. Parents aggressively defend nest site. 3–9 eggs, usually 5.

Movement:
All of the northern birds migrate south in the fall, but may be present year-round in the rest of U.S. Migration as far as southern Mexico in the winter.

Interesting Fact:
May practice polygyny, in which males may have more than one mate and tend to several nests at the same time. Perhaps certain males are better providers, or there are more females present than males. The reason is not known.

Sharp-shinned Hawk

Alan Nelson

Latin Name: **Accipiter striatus**

Field Marks: **Length 10–14 inches**
Wing span 20–28 inches

■ Summer
■ Year-round
■ Winter

Smallest and most common North American accipiter, or short-winged, long-tailed forest hawk. About the size of a robin. "Sharpies" may be distinguished from the larger Cooper's Hawk by round heads and squared-off or notched tails. Females much larger than males. Adults reddish brown on the underside and blue-gray on the back, males bluer. Juveniles up to one year have brown streaks on a white underside, with brown "dorsally," or on the back. Are active in flight with quick wing beats, often with a glide after five or so flaps as they get to the cover of the woods, and fast wing beats when hunting their avian prey. Young birds have yellow eyes that change to red as the birds get older. Sharp-shinned Hawks get their name because the leading edge of leg has a pointed keel.

Habitat:
Thick stands of immature Douglas fir and other conifers during breeding season. Sharp-shins are very secretive birds.

Behavior:
May be seen briefly at bird feeders. Their diet is 90% birds caught in flight by chasing down in bursts of speed and tactic of chase and surprise. May hide in foliage and use landmarks like hills, houses, and hedges to suddenly appear and follow the one prey bird that isn't paying attention, one that is disabled in some way, or one that is just unlucky. May still-hunt from perch. Often have "butcher block" for plucking prey and removing parts that are discarded, which is usually near nest.

Nest and Eggs:
Stick nests built each year, even on top of old ones, usually on limbs right next to tree trunk. Nests lined with bark and greenery. Nests in dense conifers and difficult to locate. Look for plucked bird feathers from prey. 4–5 eggs, incubated only by female.

Movement:
Most migratory of accipiters, some birds moving far south into Central America for winter. Many winter in Northwest. May move from mountains to valley bottoms.

Interesting Fact:
Large female Sharp-shin may be mistaken for male Cooper's Hawk. They have very similar coloring, live in the same habitats, and fly with the alternating flap and glide. Although their sizes don't overlap, it is often difficult to judge size in the field, especially with such swift fliers.

Latin Name: ***Accipiter cooperii***

Field Marks: ***Length 14–20 inches***
Wing span 29–37 inches

Summer
Year-round
Winter

Our medium-sized accipiter, with short wings and a long, rounded tail and squared-off head. The larger females may be the size of a crow, and both sexes are colored like their smaller relative, the Sharp-shinned Hawk. Young of the year are brown. Flies with several hurried beats followed by brief glide. Males show off their bright white undertail feathers in courtship displays in the spring. Yellow eyes become orange, then deeper red, with age.

Habitat: Deciduous and coniferous woodlands, usually near water.

Behavior: Feeds mostly on birds and small mammals. Sits quietly, hidden in tree limbs and leaves, dashes out in quick short flights. Very maneuverable in dense foliage. May fly along fence rows to surprise prey. Very energetic in chasing prey even on foot in the brush, and even down burrows. Smaller male tends to catch the smaller, more numerous birds close to the nest to feed female and young. Later in nesting female may travel farther to find larger birds. May locate quail by listening for their call.

Nest and eggs: Large stick nest built by both sexes in the crotch of limbs at tree trunk, 20 to 60 feet high. Always lined with wood chips and bark flakes, and usually with green leaves. 4–6 eggs, incubated mostly by female. Young hatch at around 32 days and fledge, or leave the nest, after a month.

Movement: Northernmost birds migrate, though many are sedentary, or stay put. May simply move into lower altitudes for winter. In autumn females, both young and old, migrate before males. Males return first in spring.

Interesting Fact: Earlier this century these birds were considered vermin, or pests, and were killed in great numbers. A bounty was even offered and hunters were given money for killing these "chicken hawks." Their carcasses were tacked on barns and spread on fences to "teach other hawks a lesson." Today these birds, as with all native bird species, are very much protected by state and federal laws.

Northern Goshawk

Milo Burcham

Latin Name: **Accipiter gentilis**

Field Marks: **Length   21–26 inches**

Wing span   40–46 inches

■ **Summer**

■ **Year-round**

The largest of the accipiters. Females the size of a Red-tailed Hawk. Adult plumage of both sexes very distinctive with fine gray bars on the breast and belly, slate gray backs, and black bands on the long tail. Juveniles brown. All ages have white eyebrow or "supercilliary line" over eye that may distinguish them from smaller Cooper's Hawk. May soar each day, especially the males during breeding season showing off territory, but usually secretive birds. Females especially defensive around nest and may attack people.

Habitat: Mature and old-growth forest, usually coniferous, but can be in deciduous and mixed. Remote wilderness areas in northern U.S. Prefers openings, clearings, and forest edges for hunting.

Behavior: Wide variety of prey. Birds of all sizes including grouse, and mammals like ground squirrels, tree squirrels, and snowshoe hares. May hunt from a perch or cruise forest and forest edge. Prey mainly caught on ground.

Nest and eggs: Nests near trunk in tall trees. Bulky stick nest often used year after year. Uses more green limbs and sprigs than any other North American raptor. 3–4 eggs.

Movement: Usually stays in hunting areas year-round. Goshawks in the far northern Canada and Alaska migrate south in October and November and return in spring.

Interesting Fact: The snowshoe hare is an important prey. It goes through cycles about every 10 years where the population steadily rises, then crashes. Goshawk breeding success also declines with the loss of hares, and their movements into the southern U.S. in the winter in search of food are also in these 10-year cycles, demonstrating how one species may be affected by another.

Red-tailed Hawk

Latin Name: **Buteo jamaicensis**
Field Marks: **Length 22 inches**

Wing span 50 inches

Summer
Year-round
Winter

Our most common buteo, or soaring, open-country hawk. Usually lighter above with a darker brown belly. Reddish tail occurs only in adult birds at the age of 2½; before that tail is brown with many dark bars. Subspecies are light-colored "Krider's," which breeds in Northern Great Plains, and very dark brown "Harlan's," which breeds in Alaska. Call a very distinctive, harsh "keeeeer" descending in tone.

Habitat:
" Ubiquitous," which means everywhere: countryside, deserts, field and farmlands, and woodlands in North America. May favor forest edges.

Behavior:
Hunting tactics range from "sit and wait," where Red-tails perch out in open on phone poles and snags, to soaring and scanning method of catching updrafts or thermals to observe large hunting area. Feeds on small to medium-sized mammals and birds, plus reptiles like snakes.

Nest and eggs:
Sturdy stick and twig nests built in trees or on a cliff, often lined with bark, sprigs of evergreen and fresh green foliage. May be used year after year. Nests may be used by other birds like Great Horned Owls. 2–3 eggs. Youngsters fledge at about 45 days.

Movement:
Often year-round residents that have found a steady food supply, especially during mild winters. Winter birds may be northern migrants that have come this far south.

Interesting Fact:
May be one raptor species that has profited from human presence. Forest clearing and agriculture practices have ensured the Red-tail a spot in successful breeding niches. A very abundant and thriving species.

Milo Burcham

Latin Name: **Buteo swainsoni**

Field Marks: **Length 21 inches**
Wing span 52 inches

■ Summer
■ Winter

Buteo found mostly east of continental divide, and only during summer. Large soaring hawk. Adults with brown "bib" or upper chest and white throat patch, light or barred below. Immatures spotted without bib. Tail long and light gray brown with bars and dark band at tip.

Habitat:
Bird of the Northern Plains of the West, arid and semi-arid country. Often seen on fence posts and even on ground.

Behavior:
Unusual in that diet varies with season. May hunt rodents and some small birds during the breeding season, switching to a diet of insects like grasshoppers, locusts, and dragonflies in the winter. Prey usually taken in flights from perches. Pasturelands and hayfields provide good foraging country.

Nest and eggs:
Flimsy stick nest usually built in solitary tree, shrub, or shelter belt. Nest often lined with green vegetation. 2–3 eggs, incubated mostly by female.

Movement:
Among the most migratory of North American raptors. Entire breeding population moves to Argentina, South America, to overwinter using mostly soaring and gliding as the power for their journey. Migrates only across land. As many as 400,000 seen in Veracruz, Mexico, pooled up to avoid ocean crossing. May feed along the way. Total distance in 8 weeks of migration exceeds 7,000 miles one way.

Interesting Fact:
Organochlorine pesticides recently killed tens of thousands of Swainson's Hawks on their wintering grounds in Argentina. These poisons were used to kill the insects that these birds feed upon. These pesticides are now banned in the United States, but not in South America.

Latin Name: **Buteo lagopus**

Field Marks: **Length 22 inches**

Wing span 56 inches

Summer

Winter

Large wintertime hawk. Many plumage variations between young and old, male and female. Generally adult female and young have dark belly band. Adult male has dark chest and lighter below. All have white at base of tail and dark tip. In flight, dark wrist patches show on wings. Called "rough-legged" because feathers extend all way down to toes. Species name *lagopus* means "hare foot."

Habitat:
Open country fields and marshes. Easily spotted on telephone poles and fence posts.

Behavior:
Feeds on 90% small mammals—lemmings in the north, voles in winter. Voles are small rodents like mice. Uses vantage point to find prey rather than soar like Red-tail because thermals don't occur in cold winter weather. Sometimes hovers like a huge, slow-winged hummingbird over fields looking for food.

Nest and eggs:
Nests in northern tundra where there are no trees. Builds bulky stick nests lined with grass, hair, and feathers on cliffs and sometimes hill slopes. Eggs 3–5 in number and may be as many as 7 when lemming numbers are high. More food available means more young.

Movement:
Entire population of these arctic tundra breeders moves into Canada and U.S. for the winter. Migrates in valleys at lower altitudes than many other hawks. Many Red-tailed Hawks leave the area in the winter, so Rough-legs replace them, eating similar diet. In Europe and Asia this same species is called a Buzzard.

Interesting Fact:
In winter Rough-legs may roost communally, spending the nights in one place in groups numbering up to several hundred. It may be that birds congregate in areas that offer warmer temperatures throughout the night. This is called a "favorable microclimate".

Erwin and Peggy Bauer

Latin Name: **Buteo regalis**

Field Marks: **Length 23 inches**

Wing span 53 inches

Summer

Year-round

Winter

The largest soaring buteo hawk. Pale and dark morphs, or color forms. Adults rufous above and pale below. Leg feathers also reddish, forming a V on belly when seen from below. Wings held in a dihedral when soaring and are fairly pointed. Young lighter with whitish leg feathers. Large head and gape, or wide mouth opening. Legs feathered to toes. Drop in population and local disappearances have led to requests that they be listed as threatened species.

Habitat:
Semiarid open country, prairie, desert and grassland. Also agricultural land, especially in winter. Species of the western continental U.S. only.

Behavior:
Feeds on most readily available mammals such as rabbits, ground squirrels, pocket gophers, and prairie dogs. Searches for prey in high soaring flight, or from a vantage point often on an outcrop on ground. Also quarters back and forth in flight over hunting grounds, and hovers. May eat large birds like grouse, particularly in winter when many rodents hibernate.

Nest and eggs:
Huge nests built on tops of trees, cliffs, power poles, and even on ground and haystacks. Typically in lone tree or on hilltop with view in all directions. Use large sticks and branches. Usually 3–4 eggs. Smaller males fledge earlier than females. Leave nest at 38 to 50 days after hatching. Very sensitive to disturbance, and may abandon nest. Will use and benefits from artificial nest platforms.

Movement:
Northern birds move south. Many remain in plains states year-round.

Interesting Fact:
Historic partner with the formerly huge numbers of wild bison, or buffalo, of the Great Plains. Ferruginous Hawks built nests using bison bones and lined them with bison dung. May still use cow pies to decorate nests. Called "Spotted Eagles" by Native American plains people for whom they hold great cultural importance.

Latin Name: **Aquila chrysaetos**

Field Marks: **Length 30–40 inches**
Wing span 80–88 inches

- Summer
- Year-round
- Winter

Large raptor with seven-foot wingspan. Dark brown overall with golden feathers on back of head. Young show white patches in middle of wings and white at base of tail up until breeding age of 4–6 years. Adult males have several light bars on tail, females one band. Called "booted eagle" because feathers reach down to the toes. In flight holds wings in slight dihedral, versus flatter profile of Bald Eagles. Six subspecies across Northern Hemisphere.

Habitat:
Open, often remote country. Mountains, plains, plateaus, deserts with winds and thermals to help hunting abilities. Less successful where thick vegetation and shrubs provide cover for prey and obstruct low flight.

Behavior:
Prey is mostly medium and large mammals: rabbits, hares, ground squirrels, marmots, and young deer and antelope. Birds such as ptarmigan and pheasant taken especially in winter. Also carrion. Hunts from high perch or cruises low over terrain.

Capable of high speed chase and falcon-like stoop. Pairs may hunt together.

Nest and eggs:
Large stick nests on cliffs, lined with greenery. In some regions nests in trees. Nests may be used year after year by mated pair. 1–3 eggs, incubation 41–45 days. Often only one chick survives. Fledging 65–80 days. Young dependant upon adults for several months afterward. May live up to 38 years in wild.

Movement:
Northern birds may migrate into Mexico and Central America. Elsewhere they are probably year-round residents.

Interesting Fact:
More than 20,000 Golden Eagles were killed in the southwest U.S. in the 1950s and '60s because they were known to kill newborn sheep. With this and the killing of coyotes, the rabbit population increased to numbers that reduced the forage of grass required by the sheep. Golden Eagles have been legally protected in North America since 1970.

Bald Eagle

Milo Burcham

Latin Name: **Haliaeetus leucocephalus**

Field Marks: **Length 31–37 inches**
Wing span 70–90 inches

■ Summer
■ Year-round
■ Winter

A huge raptor and the national emblem of the United States. Adult birds unmistakable with white heads (not really bald) and white tails with very dark bodies, large yellow beaks and yellow eyes. Immatures vary according to age and only have white head and tail after 4 or 5 years. In meantime plumage shows white patches throughout with a dark beak and eye. In flight tends to hold wings flat straight across. Southern Bald Eagles considered endangered until 1999.

Habitat:
Usually around bodies of water like rivers, lakes, and reservoirs. Open country.

Behavior:
Feeds mostly on fish, but aggressive predator that is able to kill waterfowl and medium-sized mammals. May steal prey from other raptors like Ospreys and other Bald Eagles, which is called "pirating", or "kleptoparasitism." In migration and over winter eats a lot of carrion, especially dead salmon and roadkills.

Nest and eggs:
Huge stick nests built in large snags or broken-off tree tops, often used and added to year after year by parent birds that may mate for life. Some nests may be 8 to 10 feet across and weigh many hundreds of pounds. Usually lay 2 eggs with both parents incubating for 35 days. Young fledge at about 2½ months.

Movement:
Northern birds from Alaska and Canada migrate south in winter to the continental U.S. Resident birds may be present year-round, as well as northern migrants.

Interesting Fact:
In Alaska huge numbers of Bald Eagles take advantage of salmon spawning runs, in which the adult fish die after breeding. This provides a very rich food source and easy pickings for hundreds of eagles during fall migration. These birds congregate in large numbers and all get along for a while.

American Kestrel

Latin Name: **Falco sparverius**

Field Marks: **Length 10 ½ inches**
Wing span 23 inches

Summer
Year-round
Winter

Smallest and most common falcon in U.S. Light colored with spots or streaks below, brown above. Males have slate gray "shoulders," or wing coverts, and solid brown tail with black bar at tip. Female has barred tail, and both sexes have two black stripes on each side of face. Immatures look like adults, with more streaking below. Very conspicuous and noisy, seen perched on telephone lines bobbing tail and head. Call is a sharp "killy, killy, killy." Probably the most abundant daytime raptor on continent. Used to be called Sparrow Hawk, but "kestrel" indicates falcon and is the name given to this type of bird around the world.

Habitat:
Open country, farmland, orchards, and fields throughout North America.

Behavior:
Feeds mostly on large insects like grasshoppers, plus rodents such as mice and voles, with the occasional catch of small birds. May hover with rapidly beating wings when hunting. Family groups are seen hunting together in the late summer.

Nest and eggs:
Falcons do not build nests, and kestrels are no exception. Woodpecker holes or natural cavities in trees are favored sites. May take over old magpie nests. Eggs number 3–5. Incubation is by both parents, but mostly the female for 28–31 days.

Movement:
We may enjoy kestrels year-round during mild winters, but usually these birds migrate to the southern states and even into Mexico and Central America. May migrate in small groups of three or four birds.

Interesting Fact:
By building nest boxes, one may attract a family of these birds. Place box about 15 inches tall with an 8 by 8-inch floor and a 3-inch hole overlooking a hunting area with some good nearby perches. With luck, some kestrels will discover this human-made home, which may be occupied year after year.

Latin Name: **Falco columbarius**

Field Marks: **Length 12 inches**
Wing span 25 inches

Summer
Year-round
Winter

A stocky little falcon, slightly larger than more common kestrel. Males blue gray above, females and immatures brown. Darker tail with light bands. Underparts streaked, throat white. Females may be larger than a pigeon, and used to be called Pigeon Hawk because they look like pigeons when they fly. Same species lives across Northern Hemishere.

Habitat:
Northern woodlands across North America (and even Europe and Asia). Open country and forest edge.

Behavior:
Usually solitary. Diet mostly birds caught in mid-air. Fast, direct flight with powerful wing beats. May hunt from perches or cruise low to the ground using trees, shrubs, hills, even houses as cover to dart out and grab prey with feet. May pluck prey at favorite spot. Will chase away larger raptors year-round, not just in nesting season. May cache food, or store uneaten portion in a hiding spot, to return to later.

Nest and eggs:
Uses old stick nests of Black-billed Magpie, sometimes hawk or crow. 4–5 eggs. Male does nearly all hunting for female and young. Young leave nest at one month.

Movement:
Three subspecies in North America, and Canadian or Taiga Merlin the most migratory, moving south into Central and South America. Some Prairie Merlins stay all year in cities eating sparrows. Black Merlin of Pacific coast is generally nonmigratory.

Interesting Fact:
Merlins are attracted to bird feeders, especially when migrating and over winter. Feeders are like smorgasbords or buffets to these birds and allow them to choose prey, often weak members of the flock. Especially fond of Bohemian Waxwings that move into western valleys in huge numbers for the winter.

Latin Name: **Falco mexicanus**

Field Marks: **Length 15½ –19½ inches**

Wing span 35–43 inches

■ Summer
■ Winter

Large, pale brown falcon of open country, with rusty upperparts and light underneath with dark spots on adult, and streaks or bars on immature. Dark axillar, or armpit, feathers tell a Prairie from a Peregrine; both are about the same size. On head, a white stripe above the eye and one behind the mustache stripe mean Prairie.

Habitat: The arid West, and open country with cliffs for nesting.

Behavior: These birds are experts in both ground and aerial prey. Will take on abundant ground squirrels all summer, then switch to birds in the air, such as Horned Larks, in winter months. Low flying tactics surprise both.

Nest and eggs: On a cliff or ledge; look for the "whitewash," or bird feces, to mark the exact location of a falcon "eyrie" or nest. Will make a scrape or indentation on a ledge to lay 4–5 eggs. Occasionally uses old common raven or Golden Eagle nests, and maybe a high cutbank in a waterway.

Movement: Young move in all directions—north, south, east, west—after being kicked out of breeding grounds by parent birds. Winter hunting ranges are often quite large.

Interesting Fact: Prairie Falcons did not suffer the same population declines as the result of DDT poisoning as their cousins the Peregrines. Birds and fish hold more poisons than mammals due to biomagnification, the tendency for the concentration of toxic substances to increase as one moves up the food chain. Rodents eat vegetation (short food chain), whereas many birds feed on insects that have eaten other invertebrates and so on (long food chain). Prairie Falcons were spared due to their largely mammalian diet of ground squirrels all summer. But Prairie Falcons switch to a largely bird diet over the winter after the squirrels have hibernated.

Latin Name: **Falco peregrinus**

Field Marks: **Length 16–20 inches**
Wing span 36–44 inches

■ Summer
■ Year-round
■ Winter

Among the most famous raptors with 19 sub-species across the world and 3 in North America. Cosmopolitan. Sexes alike in plumage. Large, stocky falcon with dark hood and mustache, or "malar" stripe. Bluish black above with spotted light underparts with rufous hue. Juveniles are browner. Skin around eye (eye-ring) and at base of beak (cere) changes from blue to yellow after one year. Used to be called Duck Hawk.

Habitat:
Open country, cliffs, and recently even large cities, where they nest on window ledges and feed on pigeons. One of most widely distributed birds in world.

Behavior:
Primarily feeds on birds. Peregrines are one of the fastest flying of birds, with stoops or dives at over 200 miles per hour. Prey are hit in the air and knocked to the ground with falcon's closed feet, or may be plucked from the air in high-speed chase.

Nest and eggs:
Cliff nesters with only a "scrape" or bare area needed to lay the typically 3 or 4 eggs. Also skyscraper ledges and bridges. Young hatch at 32–35 days and learn to hunt with parents.

Movement:
Name *peregrinus* means "wanderer," referring to this falcon's long-distance migrations. Subspecies from Canada (*Falco peregrinus tundrius*) the most migratory. Our continental subspecies, *Falco peregrinus anatum,* migrates into South America in the winter.

Interesting Fact:
Seriously endangered after the widespread use of DDT from the 1940s until 1972. This pesticide was used to control insect pests on crops, and caused eggshell thinning so that reproduction failed. Species was extinct east of the Great Plains. Over 6,000 Peregrines bred in captivity have been reintroduced to the wild by Peregrine Fund. Gradually the original numbers have been restored. Continental subspecies removed from Endangered Species List in August, 1999.

Latin Name: **Falco rusticolus**

Field Marks: **Length 20–25 inches
Wing span 50–64 inches**

■ Summer
■ Winter

Largest falcon in world. Three color morphs, gray being the most common, plus white and nearly black. Young with brown markings and bluish skin around eye, beak, and feet. Adults may be gray with dark markings, with yellow skin and feet. Relatively long tail. About the size of a Red-tailed Hawk. The national bird of Iceland. Circumpolar, or living in arctic regions across North America, Eurasia, Greenland.

Habitat: Rare winter visitor from arctic tundra. Open country such as farmland.

Behavior: Feeds on birds like grouse, pheasants and ducks, often taken on ground and water. May scan countryside from power poles. Powerful, swift, and direct flight. May fly up and take prey from above in short stoop from above. A spectacular flier.

Nest and eggs: Like other large falcons, nests on cliffs with some overhanging protection, relatively safe from mammal predators. Sometimes uses stick cliff nests of Common Ravens, Rough-legged Hawks, and Golden Eagles. Usually 3–4 eggs.

Movement: Migrants are mostly young first-year birds and adult females. Males sometimes stay on breeding ground over winter and hunt in arctic darkness. Only reach the very northern edge of lower 48 states, but some years make it to central U.S.

Interesting Fact: Gyrfalcons have long history in the 4,000-year-old sport of falconry. During the Middle Ages in Europe, only kings could keep a Gyr. A Gyrfalcon in the white phase was a sign of importance or status. Today birds are often bred in captivity. Sometimes crossed with Peregrine Falcons creating a hybrid. These resulting hybrids are valued by falconers for having the best qualities of each species for hunting . Larger numbers of young may be produced by removing the first clutch of eggs. The female will often lay a replacement clutch after about two weeks.

Milo Burcham

Latin Name: **Pandion haliaetus**

Field Marks: **Length 22–25 inches**
Wing span 58–72 inches

Summer
Year-round
Winter

The "Fish Hawk" that lives nearly everywhere in the world where water occurs. Large, long-winged, brown above and creamy white below. White head with dark stripe through eye. Female usually has neckace of dark feathers. Immature has buff-colored edges to feathers. In flight wings bend back at wrist and look like two shallow arches when soaring. Called the "citizen of the world."

Habitat:
Rivers, lakes, ponds, reservoirs, and marshes. Accustomed to people, so even in urban areas. Anywhere that fish may be found near the surface.

Behavior:
Feeds almost entirely on live fish, catching them in a headfirst dive, then swinging out legs, often going completely underwater. Outer toe moves behind to grab fish with two toes in front, two behind. Scales on toes are spikey with "spicules" to better grasp slippery prey. Often hovers, or flaps to stay in place and locate prey. Prefers "rough fish" like suckers and squaw fish. Male provides all food for brooding female and young.

Nest and eggs:
Large stick nests on snags, power poles, and human-made platforms. Material added year after year by males breaking off dead limbs in flight from standing trees. May decorate nest with harmful material like fishing netting, baling twine. Nearly always two young are raised, rarely three. Power pole nests are safe from ground predators, but not from Great Horned Owls.

Movement:
Long-distance migrant. First adults, then youngsters fly south in September. Will fly across large bodies of water, unlike other raptors. Winter in southern Central and northern South America. Young remain near wintering grounds until old enough to breed at three years. Adults return in spring to same nest, and young eventually show up in area in which they were raised.

Interesting Fact:
A single species, the Osprey, lives all around the world (except Antarctica). Next to the Peregrine Falcon this is the most widely distributed diurnal raptor. Were thought to be in a family of their own, now considered by many to be a subfamily of the Accipitridae.

Milo Burcham

Latin Name: **Cathartes aura**

Field Marks: **Length 27 inches**
Wing span 69 inches

☐ **Summer**
☐ **Winter**

Large brownish black bird with head and neck bare of feathers. Head on adults red, juveniles gray. In flight holds wings in dihedral, or upward in a shallow V. Usually seen soaring, or sometimes on perches like fence posts, holding wings open to the morning sun.

Although included here in traditional order, due to recent DNA research Turkey Vultures are no longer considered raptors, but instead usually classified in Order Ciconiiformes, along with storks and herons.

Habitat:
Open country such as farmland and deserts, and woodlands bordering such habitat. Common in north only during warmer months because only then are thermals, or updrafts, present, and food isn't frozen.

Behavior:
Feeds almost entirely on carrion, or dead animals, usually medium- and large-sized mammals. Head is bare of feathers for this reason—to keep clean. Very tolerant of toxic bacteria that other

animals won't eat, but if meat is too rotten will not feed. May roost in large groups.

Nest and eggs:
Uses caves, cliffs, and hollow logs or stumps with little or no nest material added. Sometimes on ground in dense undergrowth. Usually 2 eggs laid. Parent birds regurgitate food from the crop to feed young, unlike true raptors.

Movement:
Northern and western birds migratory, some travelling to South America.

Interesting Fact:
Turkey Vultures are among the few birds with a well-developed sense of smell. May locate carcasses from the odor. Gas pipeline companies may locate a break in the line by pumping in a bad smelling chemical, then watching where the Turkey Vultures collect, so these birds are helpful in these situations. More importantly, they clean up the wildlands and roadways. These vultures were mistakenly called "buzzards," a term that more accurately refers to buteo hawks in the Eastern Hemisphere.

Order Strigiformes

Order Strigiformes (Owls)

- **Family Tytonidae** (Barn Owls)
  - Barn Owl

- **Family Strigidae** (Typical Owls)
  - Flammulated Owl
  - Western Screech-Owl
  - Great Horned Owl
  - Snowy Owl
  - Northern Hawk Owl
  - Northern Pygmy-Owl
  - Burrowing Owl
  - Barred Owl
  - Great Gray Owl
  - Long-eared Owl
  - Short-eared Owl
  - Boreal Owl
  - Northern Saw-whet Owl

Alan Nelson

Latin Name: **Tyto alba**

Field Marks: **Length 16 inches**
Wing span 42 inches

- Summer
- Year-round
- Winter

Easily recognized owl that lives around the world with many races or subspecies (perhaps up to 46). Large head with heart-shaped facial disk and small, close-set dark eyes. Slim body with long legs and strong talons. Buff brown above with creamy white underparts. Often described as "ghostlike" in flight. Females slightly darker than males.

Habitat:
Countryside with open fields, hedgerows, farmland, and semiarid lands, especially with old outbuildings.

Behavior:
Diet almost entirely small mammals: mice, voles, pocket gophers, rats, shrews. With excellent hearing, can catch prey by sound on the darkest of nights. Very nocturnal, and usually hunts from perches.

Vocalization:
Most vocal during breeding season. Diverse calls include screeches, hisses, twitters, yelps, screams, and wheezes, especially by male. Like other owls, clicks beak and spreads wings in defense.

Nest and eggs:
Called Barn Owl because of preferred nest and roost sites in old buildings. Also nests in hollow trees and caves. Lays 4–7 eggs, sometimes as many as 12 when prey plentiful. Male feeds female and young up to about 2 weeks, then female may hunt as well. Uses artificial nest boxes.

Movement:
Very rare and local in northern part of range, probably because of harsh winters. Young especially move south in winter.

Interesting Fact:
It is said that a single Barn Owl with a ten-year life span may eat 11,000 rodents. These rodents might otherwise spread disease, foul human food, and may eat the equivalent of 13 tons of crops. Clearly, the Barn Owl, like so many other raptors, is the farmer's best friend.

Judy Hoy

Latin Name: **Otus flammeolus**

Field Marks: **Length 6 ¾ inches**
Wing span 16 inches

Summer
Year-round
Winter

Tiny owl that is rarely seen. Gray brown plumage above and lighter below with fine dark streaks. Small ear tufts on flattish head with dark brown eyes and gray brown beak.

Habitat:
Open coniferous forest in mountains, ponderosa pine, Douglas fir, sometimes with associated aspen groves. May hunt in selective logging areas.

Behavior:
Feeds almost entirely on insects and other invertebrates that are caught in the air or plucked from leaves: moths, crickets, grasshoppers, and spiders. Mainly nocturnal, but can be crepuscular, especially in breeding season. Flies out from perches in hunting area. May catch insects on ground.

Vocalization:
Perhaps the easiest way to locate "Flams." A deep short "hoop" repeated every 2–3 seconds, or maybe "hoo-hoop" when excited.

Nest and eggs:
Cavity nesters taking advantage of woodpecker holes usually 15–40 feet above ground. Eggs number 2–4, sometimes 5. Incubation short, only 21–24 days, entirely by female.

Movement:
May be the most migratory of North American owls, because food supply of insects disappears in winter. Northern birds move to Mexico and Central America in fall, and return at lower elevations in spring to feed on insects on the way.

Interesting Fact:
Flammulated Owls were once considered quite rare. Their feathers are colored exactly like the bark of a tree, so they blend in exactly with their surroundings. We call this cryptic coloration. They also have soft voices and are active only in low light, so they may be more common than we think.

Latin Name: **Otus kennicottii**

Field Marks: **Length 8¹/₂ inches**
Wing span 20 inches

Year-round

Jay-sized owl that is locally common, or in isolated pockets. Gray above and below with perfect camouflage, or cryptic coloration of streaks and spots that hides them when perched next to the trunk of a tree. Ear tufts raised when roosting or alarmed, but tufts may lie back. Yellow eyes and gray bill. Very similar to Eastern Screech-Owl, which has red color morph. Eastern species occurs east of continental divide, especially along Missouri and Yellowstone rivers in Montana.

Habitat:
Woodland and forest edges, semi-open country, city parks.

Behavior:
Feeds on small rodents and birds. Also insects in warmer months—crickets, beetles, and moths—that are largely active at night like screech-owls. Pounces on prey from different perches. Roosts in day next to bark of tree.

Vocalization:
Really doesn't "screech" at all. Series of low-pitched whistles speeding up like a bouncing ball coming to a standstill. Short trill followed by longer trill. Mated pair may sing a duet together, male calling then female, back and forth and often overlapping.

Nest and eggs:
Tree-cavity nesters— woodpecker holes, especially flickers, very rarely in magpie nests. Eggs 3–7, young hatch at 26 days, fledge at a month. Fed by parents for 5–6 weeks. May roost in cavity during winter.

Movement:
Probably don't move other than young dispersing in fall. Nonmigratory.

Interesting Fact:
Eastern and Western Screech-Owls were considered the same species until the 1980s. Differences in vocalizations, size, and genetics have proved that the two forms are actually different species that rarely interbreed.

Latin Name: **Bubo virginianus**

Field Marks: **Length 22 inches**
Wing span 44 inches

Year-round

Large powerful owl. Can vary in color from rich dark brown to sandy tan in arid country. Bulky, with ear tufts that may help in camouflage when roosting. But ear tufts may be lowered, making head look flat. White throat area with dark patches below. Yellow eyes with bluish gray beak. Facial disk light to dark orange.

Habitat:
Woodlands, open country, farmland, desert, mountains, swamps—nearly everywhere except very dense forest and treeless prairie. No other owl in North America lives in so many habitats and climates.

Behavior:
Generalist in feeding. Always takes advantage of easy prey, no matter what the size. Feeds on earthworms, insects, crayfish, fish, reptiles, amphibians, and birds. Large portion of diet is mammals, from shrews, mice, voles, on up to squirrels, rabbits and hares, even skunks and porcupines. Kills many other species of owls and hawks. Mostly nocturnal, hunting from high perches. Mobbed by crows and other songbirds if seen during day. Small birds are warning others that a predator is in area.

Vocalization:
Loud, deep hoots in five-seven syllable series that may be like "Don't Kill Owls, Save Owls." Male more vocal, fewer hoots with deeper voice than female. People can attract Great Horned Owl by imitating call in late winter where they breed.

Nest and eggs:
Nests as early as January. Usually has several nest sites from which to choose. Old broken-off snags, stick nests of hawks, crows, herons, sometimes rocky cliffs. 1–3 eggs, usually 2. May pair for life, but will accept another mate if one is killed. Young kicked out of territory in fall by adults, and must seek their own hunting grounds.

Movement:
Nonmigratory and remain on territory year-round.

Interesting Fact:
Most common owl in North America, and also largest by weight aside from the Snowy Owl, which doesn't actually breed in the lower 48 states. May be most successful of our raptors, perhaps due to their generalist diet and tolerance of people.

Gary Lackie

Latin Name: **Nyctea scandiaca**

Field Marks: **Length 23 inches**
Wing span 52 inches

■ Summer
■ Year-round
■ Winter

Unmistakable. Huge white owl from the arctic, present in lower 48 states in winter. Female larger with fine dark bars to better blend in to surroundings on the tundra while nesting. Male all white, or with some dark bars on wings, tail, and/or back. Eyes relatively small and yellow, beak black. Head large and round, and body feathers very dense and heat retaining. First-year birds darker because of heavy barring, males with browner markings. Hawklike in flight.

Habitat:
Arctic tundra from northernmost reaches of land across Northern Hemisphere south to tree line in summer. Areas with low vegetation, hills, rock outcrops. In winter may venture into southern Canada and northern U.S., especially in "irruptive years" of low food supply. Then may be seen in agricultural land, airfields, marshes, prairies perched on ground, rock, or fence post.

Behavior:
Diurnal and crepuscular. Feeds nearly entirely on lemmings (arctic rodents) over summer. Also voles, rabbits, hares, ground squirrels. In winter may take more birds, up to the size of a goose. Usually hunts from low perch with swift, direct flight. Locates lemmings under snow using hearing. May be easily approachable by humans.

Vocalization:
Usually silent, except around nest when breeding. Male has harsh, grating bark, and a deep, low hoot. Female call is higher pitched.

Nest and eggs:
Nest on ground, usually on small hill, mound, or rock outcrop to afford view of ground predators like arctic foxes. Clutch size varies with prey availability, and may not nest at all when lemming populations crash. Usually 3–5 eggs, sometimes up to 11. Young may wander from nest on foot at two weeks, leave nest at 20–28 days, fly well at 50 days. Cared for by both parents up to 10 weeks afterwards

Movement:
Migratory and nomadic. Some remain on breeding grounds when food permits. Numbers in continental U.S. vary year to year.

Interesting Fact:
Largest owl in North America. Hawklike in habits and lacking the noise-reducing plumage of nocturnal owls. Body feathers dense and heat retaining. Very long feathers on powerful feet, with talons partly covered. Plumage type and coloring all very adaptive to arctic environment.

Milo Burcham

Latin Name: **Surnia ulula**

Field Marks: **Length 16 inches**
Wingspan 28 inches

■ Summer
■ Winter

In northernmost continental U.S. only in winter, but with some breeding records in Montana. Distinctive shape and posture. Medium-sized with long tail, broad, square head with black markings outlining facial disk. Brown above, white with brown bars below. Perch upright and behave like a hawk. Often seen on topmost limbs of tree such as spruce or snag, sometimes with tail cocked up at angle. Wings pointed like falcon in flight. Often easily approached by people.

Habitat:
Northern forest, coniferous and deciduous with clearings for hunting, such as bogs, burned areas, and logged units. May overwinter in farmland and prairies, perching on haystacks and fence posts.

Behavior:
Mostly hunts during the day, and at dawn and dusk. Mainly uses sight to locate prey, but uses hearing in deep snow. In breeding season feeds almost entirely on voles, with some snowshoe hares, squirrels, and small- to medium-sized birds. More birds in winter. Uses favorite hunting perches to swoop low and fast. May hover.

Vocalization:
Display call a bubbling trill lasting ten seconds or more, which may be imitated using referee's whistle. Also series of whistles.

Nest and eggs:
Nests in cavity, or broken hollow top of tree. Sometimes stick nest of hawk or crow. 6–10 eggs, larger number in years when vole populations high. Young may stay with parents for several months after fledging.

Movement:
No real migration, but nomadic like Great Gray. When vole numbers crash every 4 or 5 years, Hawk Owls, especially juvenile birds, move south for winter seeking prey. These are called years of "irruptions." Adult males may remain on breeding grounds to secure the scarce nesting cavities.

Interesting Fact:
Northern Hawk Owl may benefit from logging and fires. Habitat improves with small units logged over time, associated with patches of forest. By leaving snags and stumps, these areas may be occupied year-round. Hawk Owls seem to prefer recent burns.

Milo Burcham

Latin Name: **Glaucidium gnoma**

Field Marks: **Length 6¾ inches**
Wing span 12 inches

☐ Year-round

Tiny, plump, with round head and long tail, short wings. Gray and rufous color morphs, grayer in Rocky Mountains. Dark and spotted above, streaked below. Flanks, or sides, streaked with black. Tail dark with many light bars. Two black spots on back of head resemble eyes. True eyes are relatively small and yellow. Flight noisy, with fast wing beats and gliding. Often very tame and approachable.

Habitat:
Deciduous and coniferous woodlands, and forest edge. Seems to prefer openings rather than dense forest.

Behavior:
Very powerful for its size. Feeds on small to medium-sized birds, mammals, reptiles. Some large insects, depending on time of year. Can kill quail and red squirrels, both more than two times owl's weight. Diurnal and crepuscular. May fly from perch to perch to get closer to prey, then drop straight down. Flicks tail side to side. May frequent bird feeders, especially in winter. Often "caches," or hides, uneaten part of prey to return later.

Vocalization:
A long series of short hoots or whistles equally spaced at 1–2 seconds.

Nest and eggs:
Cavity nester. 2–7 eggs. Unusual for owls in that female may wait until whole clutch is laid before she starts incubating, so young may be same age. Female may start hunting for young at 9 days; fledging begins as soon as 23 days.

Movement:
Resident year-round. Movement from mountains to valleys with milder climate (and bird feeders) for winter.

Interesting Fact:
On back of head, markings may resemble a false face. Two black spots surrounded by white with a hint of a beak shape. Advantage is that predators that would want to attack pygmy-owl or steal its prey aren't sure which way the owl is looking. Owl can accentuate the "eyes" and make them look larger. The element of surprise is lost for the potential attacker.

Milo Burcham

Latin Name: **Athene cunicularia**

Field Marks: **Length 9½ inches**
Wing span 21 inches

Summer
Year-round
Winter

Long-legged, ground-dwelling owl. Brown with white spots. Round head with yellow eyes. Juveniles light below and buff on breast rather than streaked like adults. Considered a bird of "special concern," "threatened," and "endangered" in much of range because of loss of habitat, especially with drops in prairie dog populations. May benefit from artificial nests and perches.

Habitat:
Dry and treeless open country; plains, grassland, prairie, desert. Adapting to human surroundings at times at golf courses, airports, cemeteries, industrial parks, vacant lots.

Behavior:
Preys mostly on insects and small mammals. Prefers grasshoppers, crickets, moths, and beetles, plus mice and voles. Also some small birds, and reptiles and amphibians. Mostly crepuscular. Will hunt by running and hopping on ground. Also may hover over vegetation, and catch insects in air. Has a dashing flight from perch. Allopreening and head-bobbing common.

Vocalization:
Male's call is a dovelike "coo-coo." Also series of chattering "kack" notes.

Nest and eggs:
As name suggests, nests in burrows, usually made by other animals such as prairie dog, ground squirrel, marmot, badger. But capable of excavating own burrow, especially in loose soil. Eggs 6–11 with young leaving the burrow at 44 days, returning to roost. May add dried cow dung to nest chamber and entrance. It is very unusual for an owl to add nesting material. Often nests semicolonially, with up to a dozen pairs in group.

Movement:
Migratory in all of northern range.

Interesting Fact:
Young Burrowing Owls make special sound when they feel their burrow is threatened by a predator—a "rattling hiss" that sounds just like alarmed prairie rattlesnake. This may scare predators and keep them from entering the burrow. The young owls are excellent mimics of the dangerous "model," the rattlesnake.

Milo Burcham

Latin Name: **Strix varia**

Field Marks: **Length 21 inches**
Wing span 42 inches

Year-round

Rather large owl with rounded head and yellow beak. The "bars" of dark brown extend across the head and upper body, with vertical streaks below. Facial disk whitish gray with dark rings circling dark eyes. Back is an uneven pattern of brown, black, and cream, so bird blends in while roosting. A North American owl more common in the eastern U.S. Closely related to Spotted Owl. The two species occasionally hybridize where they live together in Pacific Northwest.

Habitat:
Mature stands of forest, coniferous and mixed trees, river bottoms. Dense woods with some clearings.

Behavior:
Feeds on rodents at night, even flying squirrels. Also birds, reptiles, amphibians, fish—usually prey that can be swallowed whole. Hunts from perches to pounce, but may hover over likely spot while seeking prey.

Vocalization:
Very distinctive nine-syllable hoot in two phrases that sounds like "who cooks for you, who cooks for you ALL" with the last part a gut-wrenching call. Also a series of deep hoots with the "you ALL" at the end. Very vocal, with shrieks, barks, trills, and squeaks. Young very noisy with begging calls after leaving nest.

Nest and eggs:
Nest in large, natural cavity or hollow of broken-off tree, occasionally in abandoned nest of crow or hawk. Eggs number 2-5. Probably long-lasting pair bond.

Movement:
Resident birds year-round with only youngsters leaving parent territories.

Interesting Fact:
Barred Owls have only been present in the western United States in any numbers as recently as the 1960s. They expanded their range west across British Columbia, Canada and then moved into Washington, Oregon, California, and Idaho as late the 1980s. They have moved across the Canadian woodlands north and perhaps are jumping the shelterbelt tree regions of Montana's plains to arrive in western woodlands.

Milo Burcham

Latin Name: **Strix nebulosa**

Field Marks: **Length 27 inches**
Wing span 52 inches

Summer
Winter

Huge-looking owl that is mostly feathers. Large round head, fluffy feathers, and long wings and tail give appearance of huge body size, but actually smaller by weight than Great Horned Owl. Large facial disk with dark rings and white "mustache" make yellow eyes look small and "staring." Overall gray with dark markings. Wing beats slow and deep. May be easily approached by people. Lives across Northern Hemisphere

Habitat:
Coniferous forest and mixed coniferous-deciduous forest for roosting and nesting, often aspen forest for nesting. Plus nearby meadow clearings and bogs for hunting.

Behavior:
Relatively small feet so prey is mostly rodents, perhaps 90% during breeding season. Also pocket gophers, mice, shrews, squirrels, hares and some birds, especially in winter. Nocturnal, diurnal, and crepuscular. May use treetops, snags, fence posts to listen and watch when hunting. Will fly low across openings to swoop down on prey.

Vocalization:
Booming deep hoots, lower pitched by male. Evenly spaced, becoming slightly faster and weaker at end of series. Also a softer double call, perhaps to defend territory around nest.

Nest and eggs:
Uses abandoned stick nests of other birds like raven, goshawk, or broken-off top of snag, hollow in stump. 3–5 eggs, maybe up to 9 with large vole numbers. Allopreening important in pair bonding. Female very aggressive in guarding nest after young hatch, and readily attacks intruders.

Movement:
Resident and stable in some areas and years. Will move south in times of poor prey numbers, or "irruption" years.

Interesting Fact:
The Great Gray has the largest facial disk of any owl. This makes a perfect circle that may help it to hear better than most. Using its keen hearing, will readily plunge into snow depths of up to a foot and a half head first, grabbing subsurface rodents with talons. Can break through a hard crust of snow for their prey. Will also capture pocket gophers by punching through dirt.

Latin Name: **Asio otus**

Field Marks: **Length 15 inches**
Wing span 36 inches

Summer
Year-round
Winter

Medium-sized owl with truly long ear tufts, close set but not always visible, especially in flight. Resembles Great Horned Owl, but smaller, and without white throat patch. Round facial disk and yellow eyes. Paler facial disk and underwing on male. Brown with dark streaks and bars that help them blend in with tree bark and lichens during the day. Lives all across the Northern Hemisphere.

Habitat: Open coniferous woodlands, mixed deciduous forest, and shelter belts. Relies on openings to hunt at night.

Behavior: Feeds mostly on small mammals, whatever is common in area: voles, mice, shrews, pocket gophers. Mostly nocturnal. Will fly low over hunting area and locate prey by hearing with very quiet flight, listening while in the air to drop down on rodents. Sight may be secondary on these occasions. During breeding, male will perform flight display with zigzagging and wing claps below body in flight.

Vocalization: A long series of "hoo" notes by the male during the breeding season, often lasting for up to 200 notes spaced at 2–4 second intervals. Variety of catlike calls given during defense of nest and young. Also barking alarm call, but usually silent, so may be overlooked in much of territory.

Nest and eggs: Will use stick nests of other birds, such as crows, magpies, and hawks. Eggs number 5–7. Male continues to feed the young up to six weeks after fledging.

Movement: Migratory in northern regions and very nomadic in areas where cyclic voles are main prey item.

Interesting Fact: May roost communally in winter in shelterbelts and trees. Winter roosts have been found containing as many as 80 birds in small area. Birds breeding in northern U.S. and southern Canada have been recovered in winter in central Mexico, more than 2,000 miles south of banding sites.

Latin Name: **Asio flammeus**

Field Marks: **Length 15 inches**
Wing span 38 inches

Summer

Year-round

Winter

Medium-sized, buff brown owl of open country. Cryptic coloration looks like dead grass. Round head with small ear tufts seen only in defensive posture, located near center of forehead. Yellow eyes surrounded by black. Light below with brown streaks. In flight, long, broad wings held in slight dihedral, dark patch at wrist. Low, bouncing flight, like a big moth. Low wing loading, so capable of slow, quiet flight while hunting. Male slightly paler than female. Roost on ground, or in low vegetation. Communal roosting common in winter with numbers from several to over a hundred on ground together. Present on all continents except Australia—a very successful species.

Habitat: Fields, marshes, farmland, prairie, tundra.

Behavior: Diet 95% small mammals, mostly voles, plus mice, shrews, pocket gophers, and a few small birds. Nocturnal, diurnal and crepuscular, hunting at all hours. Male performs "sky dance" over breeding territory. Will circle up and dive down, clapping wings below, visible from a great distance. Adults will feign a broken wing, and crash into brush to lure predators away from nest.

Vocalization: Usually quiet, but may "bark"

year-round when disturbed. Male "sings" in flight during courtship, a series of up to 16 "hoo-hoo-hoo..." notes.

Nest and eggs: Nest on ground where female may be hidden in grasses. Nearly unique for owls in that a nest is made, with female adding grass and downy feathers to form a shallow bowl. Eggs 5–10. Larger clutch size when vole numbers are high. Young leave nest on foot before able to fly to perhaps reduce discovery of nest by ground predators like foxes and skunks.

Movement: Dispersal of young, and nomadic prey seeking in which owls move in all directions. True migration occurs; some remain on breeding grounds when sufficient food supplies exist. Northernmost breeders always move south.

Interesting Fact: Large clutch size in times of numerous prey doesn't always mean survival of all. If prey numbers drop during breeding season, the oldest offspring may eat the youngest. With these birds there may be a two-week difference in age. This is called "siblicide" and may help ensure that some nest-mates are successful.

Alan Nelson

Latin Name: **Aegolius funereus**

Field Marks: **Length 10 inches**
Wing span 21 inches

Summer
Winter

Inconspicuous. Similar to smaller Saw-whet, but large facial disk has brown-black borders and bill is yellow. Eyes are yellow with white eyebrows. Upperparts buff brown with white spots. Creamy white underneath with brown streaks. Feet heavily feathered. Juvenile plumage dark chocolate brown with white eyebrows and mustache for first three months or longer. Females about 43% larger than males.

Habitat:
Northern coniferous and mixed forest across Northern Hemisphere. Spruce and fir with aspen and birch may be favored.

Behavior:
Feeds on small mammals—voles, mice, shrews—plus small birds and insects. Will hunt from perch to perch, zigzagging through forest to grab prey on ground. Primarily nocturnal. Roosts next to tree trunk rather than out on limbs, like Saw-whet.

Vocalization:
Male may call "staccato song" of trills that get louder in a series. May repeat for 20 minutes to several hours, as many as 4,000 times.

Nest and eggs:
Pileated Woodpecker nests are favored, also Northern Flicker. 3–6 eggs. Female usually chooses new mate and nest site each year. Man-made nest boxes used by 95% of breeding Boreals in parts of Europe.

Movement:
Resident, with male on breeding grounds year-round. Young and females may disperse in winter. May migrate when winter food supplies are low.

Interesting Fact:
The Boreal Owl has very asymmetric ear openings so that skull appears very different from left side to right. Uses sound to locate prey, taking noisy, moving animals more often then stationary, quiet ones. Captures prey that is active under snow and vegetation. Males may be polygynous, mating with more than one female in a season.

Milo Burcham

Latin Name: **Aegolius acadicus**

Field Marks: **Length 8 inches**
Wing span 17 inches

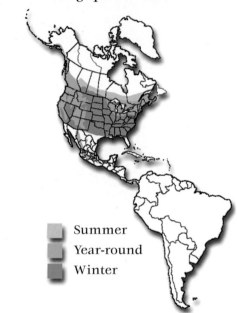

■ Summer
■ Year-round
■ Winter

Small, reddish brown above, white below with soft reddish markings. Round head with white streaks, yellow eyes, black beak. Facial disk light without black border. Juvenile red overall through the first summer into fall. When detected may pull wing forward just under beak, like a vampire's cloak. Easily approached and may even be picked up.

Habitat:
Coniferous forest. Also deciduous, mixed, streamside groves in arid areas. Requires thick cover for roosting.

Behavior:
Diet of mostly deer mice, plus voles, shrews, some insects like beetles. Small birds, especially as they roost at night. Capable of killing prey larger than itself. Nocturnal, but ocassional diurnal foraging. Hunts from low perch, often in clearing or forest edge. Roosts in dense foliage, usually around 12 feet from ground, often far out on limb rather than next to trunk. Caches uneaten portions of prey or whole prey to return to later.

Vocalization:
Quiet except during breeding season of March to May, but may call in fall as well. Male "toots" short, clear note repeated over and over again, often for hours without a break. About 130 notes per minute. Female's song softer.

Nest and eggs:
Cavity nester, favoring holes of Northern Flicker and Pileated Woodpecker. Uses nest boxes. 5–7 eggs. All feeding of female and young by male. Female leaves nest to roost elsewhere when youngest is about 18 days old. She may feed them or depart. Continued feeding by male up to a month after fledging. Young may remain together during this time. Probably no permanent pair bonding. Seldom uses same nest site two years in a row.

Movement:
Some remain on breeding ground year-round, but northern breeders are highly migratory. May move into central and southern U.S., Mexico. Others move from mountains to valleys. Migrates at night.

Interesting Fact:
Gets its name from the contact call. The "scree-awe" sound was compared to the sharpening, or whetting, of a saw blade. Audubon described it as "filing the teeth of a large saw," a sound that was far more common when this bird was named.

Raptors of the Rockies Educational Program

Wayne Tree

We have learned about the wild raptors around us. Now we will focus our attention on the Raptors of the Rockies educational program.

We have included photos and biographies of the teaching team birds, many of whom have been with us since the 1980s. These birds are actually on loan from the Federal Government, specifically the Fish and Wildlife Service. It is only with their permission, in the form of a Possession and Eagle Exhibition Permit, that the birds may be kept by us. Each bird is to be used in a minimum of twelve programs each year, and strictly for educational purposes and not for promotion of a business. Annual reports to the government keep close track of our activities.

The artwork included in the pages of this book are zinc plate etchings or pen-and-ink drawings by Kate Davis. She has created portraits of many favorite program birds by having the best of models for inspiration.

Jonathan Phillips

Kate releases a Great Horned Owl at the site where it had been retrieved three weeks earlier in August of 1992. This bird had apparently tangled with a skunk, and had been sprayed in the eye and bitten in the foot. Tube feeding, then mice doubled her weight, and we hope she learned her lesson.

Educational presentations begin with wildlife rehabilitation. We try to assist injured raptors, with the ultimate goal of returning them to their natural habitat. Occasionally a bird is unreleasable due to a permanent disability, and does well in captivity. Each individual is placed on a Federal Possession or Eagle Exhibition Permit and all are kept year-round in outdoor facilities. We also care for two Falconry Permit birds, used in discussions of the sport.

Raptors of the Rockies

Raptors of the Rockies is an educational program based in Western Montana, and has been active since 1988. Over the years, we have assembled a fine family of program birds, reached schools and public groups in hundreds of communities, and joined the space age with our website and on-line newsletters. Our move to "permanent ground" in 2001 allowed us to build all new and safe enclosures for our teaching team on the banks of the Bitterroot River. Our goal of continuing the message of awareness of the natural world around us now seems certain.

An educational program for Meadow Hill Middle School

As "chief cook and bottle washer," my responsibility is to ensure that the birds are content and in good health. That requires daily feedings of mice, chickens and quail, rabbits and squirrels, and meat of all sorts. This must be purchased, collected, butchered, and dispersed with each species' individual needs of utmost concern. The comfort of each bird is considered, with shelter from the elements, perches and roosts with astroturf to prevent foot ailments, shade cloth for cooling or heat lamps for warmth, and water baths for all. Sanitation is a constant regard. These birds are utilized only for education, and not for promotion of businesses or products. Annual reports to the Federal Fish & Wildlife Service list each program date, school or organization, contact, audience numbers and ages, and birds used in the presentation.

Program birds serve as invaluable tools in conveying the adaptations and strategies essential to raptors. Discussions center on predation, structure and function, and environmental concerns. Also included is a survey of the laws protecting wild birds, plus anecdotal tales of rehabilitation and past programs. Our hope is to educate, entertain, and illustrate the importance of the wildlife around us, plus to instill an admiration and respect for avian life and these skilled predators. A presentation of "Birds in Art" involves understanding avian anatomy and function, and sketching or sculpting sessions with the greatest of models to render in art. Our program schedule fills up with engagements through the year. We strive to increase this message of admiration for the natural world around us, and a respect for birds of prey with their beauty, grace, intelligence, and skill as predators. We are sorry, but Raptors of the Rockies is not open to the public.

Our facility in the Bitterroot Valley

Meet Our Teaching Team

Here are photos and brief bios of our raptor "ambassadors." The days listed are acquisition dates and the birds are listed from our oldest to most recent.

Rikki the American Kestrel

the **Great Horned Owl**
male December 19, 1988

Bobo the Great Horned Owl has been with the teaching team since the beginning, and has been a favorite for all of these years. He had been shot through the right wrist on the closing day of big game hunting season in 1988. The not-so-comical explanation is that someone mistook him for a spike elk with his dark ear tufts resembling antlers. He later lost the end of that wing, but still has enough insulating feathers to remain comfortable through the winter. He is very tidy, and bathes in his water pan even in chilly weather. Bobo has played the role of surrogate father to young Great Horned Owls so that they learn their vocation as an owl and not as a human before being set free. After his 500th appearance he was joined by another male Great Horned Owl to share the role of educator. Bobo and Miles are inseparable and roost side by side, often leaning against each other. With his long stage history, Bobo is recognized by children and adults alike. College students fondly recall meeting Bobo when they were "just kids." These two owls are quite vocal, perhaps responding to the hoots of wild owls. Bobo is the patriarch of the the educational birds.

Miles is the second Great Horned Owl and a welcome companion to his roommate Bobo. Miles hit a power line in the winter of 1998, and is permanently missing five of the ten primary feathers on his manus, or hand. Miles was an adult bird when he suffered his injury, yet he is very mild tempered and immediately was an excellent program bird, making his debut at the state Montana Audubon Conference. Miles probably decides who gets what as far as the six mice placed in the owl building each day, and he caches his uneaten food, usually in the form of mouse hindquarters, in various locations. Being as my last name is Davis, Miles is named for the jazz trumpet player that was a huge influence on my musical tastes, thanks to my father. A little discussion of jazz as American music may unfold at programs. Recently a 6th grader jumped to his feet to exclaim that he knew the name Miles Davis— his gym teacher in 3rd grade. Miles underwent surgery on his eye that was punctured somehow in his enclosure. This was a great cooperation between human and animal doctors, and he has recovered fully.

the **Great Horned Owl**
male April 23, 1999

the **Cooper's Hawk**
female November 18, 1989

Alice the Cooper's Hawk was the victim of a vehicle collision in the Bitterroot Valley in 1989. Probably chasing her intended meal of a flying bird, she suffered a broken right wing that failed to heal properly. Alice eats half a small chicken each day, which are supplied by a Montana hackle farmer. Hackle feathers are used in tying fly fishing flies and are the long neck plumes supplied by a rooster. The female chickens don't grow these display feathers so they are killed at five weeks. We don't feed chickens to rehabilitation birds that are to be released to prevent a bad habit from starting. Alice is named for a male rock star, and this brings up the fact that with the raptors, the females are larger than the males due to reverse sexual dimorphism. The male accipiters might be half her weight. This is a species that is usually seen in brief glimpses, sometimes at the bird feeder, which they may see as a buffet. Woodland birds, they have short wings for maneuvering through foliage and long tails for steering. Alice has been an excellent bird and displays the traits of accipiters, true hawks that are not often seen up close.

the **Western Screech-Owl**
female February 12, 1990

Crackity Jones the Western Screech-Owl, or "Jonesy" for short, has been with the educational staff since February of 1990. After a collision with a vehicle, a wing fracture at a joint healed poorly and the wing tip had to be amputated. All owls, even babies, click their top and bottom beak together as a warning, hence her name (and the name of a Pixies song). Jonesy consumes two or three lab mice daily, which may be half her body weight. In programs for kids we have a volunteer mathematician calculate how many Quarter Pounder hamburgers they would have to eat each day to be a small raptor. Students realize this would be an impossibility. Crackity Jones spends her days in the top limbs of a 3 ½ foot yew tree and calls when someone walks by. Jonesy is the best bird to demonstrate cryptic coloration, or camouflage. In 2000, she had her debut on national television on the PBS show *BirdWatch* with four of her teaching teammates. Plus Jonesy is a personal favorite and grabs a mouse with her beak, then transfers it to her foot quick as a flash. She is indeed a program professional.

Spike the Short-eared Owl was found at Ninepipe National Wildlife Refuge in December of 1990. He had been shot, and lost part of his left wing right away. Perhaps he was mistaken for a pheasant, but certainly the difference should be clear to hunters. He has been instrumental over the years as an example of the laws that protect wild birds. We might tell kids that the little boy that shot this bird AND his parents are still in prison to this day, hoping our little joke hits home. These owls are the farmer's friend as well. In ten years an owl this size may eat 10,000 mice and voles, which could destroy 13 tons of potential crops. Spike has been given the nickname "Bungie" because at some programs he leaps off the glove as if to bungie jump, and "Jiffy Pop" because sometimes he bounces in his box. His entire diet is lab mice and his pellets, or daily oral disgorgings of undigestible bones and fur, have been taken apart and examined by thousands of students over the years. Often students reconstruct the mice on cardboard using glue. Spike is excellent for displaying a predatory niche uncommon to most owls.

the **Short-eared Owl**
male December 1, 1990

Max the Golden Eagle has been in human care since a fledgling in 1989, and he joined us in Clinton in 1992. He was found in the Bob Marshall Wilderness alongside a trail unable to fly, and was packed out on horseback. His inability to fly has been attributed to some type of poisoning. In the past predators were killed outright, and perhaps Max was the victim a a poison-laced carcass. This has been an illegal practice for some time. His balance is very poor, and this will certainly never improve. He was formerly cared for at Wildlife Wildlands Institute by John Craighead. Max lives in a 30 by 40 foot fenced yard with his best friend, Nigel. They spend most of their time perched side by side or facing each other on the big round perches. Max has a beautiful call which we hear especially at sunset. Both eagles enjoy game meat of rabbit, elk, deer, and bear (and the occasional sirloin steak). Plus we buy up to 300 Columbian ground squirrels from a trapper each year, which is Max's natural diet. Max is the boss when it comes to feeding time, and Nigel gets the food on the other perch. As a youngster now, Max may be with us another thirty years. This is a major commitment, but one that we hope ensures a very valuable and charismatic member of the teaching team.

the **Golden Eagle**
male July 7, 1992

Kiko

the **Prairie Falcon**
female September 25, 1992

Kiko the Prairie Falcon is the raptor with an attitude. She was a car collision casualty as a fledgling in 1992 and still thinks that she can fly even with a healed shoulder break. She now shares her perches with Otto and Bayly, but has her own feeding platform in the new building. She tries to get her roommates all riled up, especially in the spring. Kiko has a very vocal presence, and in programs often forces participants to cover their ears lest they be deafened. Teachers have told me that this heightens the classroom experience, so we let Kiko call. Prairie Falcons feed on a largely mammalian diet of ground squirrels all summer, but switch to an avian diet over the winter after the squirrels are hibernating. This is quite a change in hunting tactics, grabbing birds in the air. Kiko eats a young chicken or quail daily, and now enjoys rabbit legs. Falcons love shelves as perches, which may mimic a cliffside hangout, and Kiko sometime lies down on her astroturf "nest." Her name comes from the Los Lobos album of the same name. She is a beautiful bird with, we hope, a very bright, albeit noisy, future.

Jallad

the **Merlin**
female February 6, 1995

Jallad is a female Merlin that has been a favorite program bird ever since her unfortunate collision with a train in February of 1995. A broken left wing did not heal, so Jallad now has a sunny building with the Screech-Owl next door. She is a migrant from Canada, a Taiga Merlin, and a representative of one of the three subspecies in North America. These birds are like little cheetahs and they will grab birds in the air in direct pursuit, like their larger relative the Gyrfalcon. Jallad eats one mouse a day, plus the occasional leg of a young chicken or quail. She spends a great deal of time lying in the sun. The name "Jallad" means "executioner" in Arabic, and as a little joke was the name of my first Kestrel in 1973 in Cincinnati. Jallad is often quite vocal, and has been an outstanding falcon for education, especially on discussions of migration and falconry.

the **Red-tailed Hawk**
female April 4, 1997

Bayly the Red-tailed Hawk is a beautiful representative of the "big hawk" familiar to everyone. She was found on a hiking trail in April of 1997. She had somehow broken her left wing, and is our only bird that had a natural injury, and not one related to people. That summer she molted into the reddish tail feathers that mean $2\frac{1}{2}$ years and adult-hood, so we know her age. Bayly has quite the mind of her own, and displays the intensity and potential ferocity of the birds of prey, traits very worthy of respect. In other words, she scares very young kids. Still, she has adapted favorably to captivity and is content, enjoying her diet of mice, heart and game meat, and squirrels. In the past she has been a model "mom" to many young red-tails, and we hope they have learned that people are to be feared to avoid the problem of imprinting, when wild animals consider people as their friends. Bayly's name comes from the surname of my aunt and uncle, who are not at all ferocious. She adds a new aspect to education by offering insight into the predatory attitude. Bayly the Bad Girl keeps us on our toes.

the **Rough-legged Hawk**
male May 20, 1997

Otto the Rough-legged Hawk has been with the Raptors of the Rockies since the winter of 1997. This youngster was found in the Flathead Valley after being hit by a car, and a broken wrist healed with an immovable joint. Otto was typical of the first-year Rough-legs with his light upper body and dark belly band. Now with several molts he is an adult male with the opposite—darker above and light below. Only in the lower 48 states during the winter, a large number of Rough-legs communally roost near where this bird was found, foraging singly to hunt during the day. Otto lives in an airy outdoor enclosure with shade cloth to keep him cool. His roommates are Kiko the Prairie Falcon and Bayly the Red-tail, and he is the calmest of the group. He feasts on mice, squirrels, chicken legs, and heart and game meat. Otto is an excellent model for our many demonstrations on Birds in Art. We sketch or sculpt some of the raptors, including this hawk and an owl and eagle—and what fitting subjects. Otto is named after a friend of our family, and not Auto, as some students think.

the **Northern Harrier**
female July 23, 1989

Mika Hawkinen, the Northern Harrier has a most interesting history. As a ground nesting species these birds are particularly vulnerable, and Mika's nest was run over by a hay swather in July of '98. The rancher retrieved this only nest survivor and called us immediately. His quick thinking saved this bird, as the front three toes had been amputated in the accident, plus a laceration to the crop threatened her life. A veterinarian sewed up the crop, then more heroics began. With friends at a local hospital we started a regime of homeopathy and human medication so this baby bird would learn to stand on a foot with a single hind toe. After a month of dressings and bandages, her foot healed so well she pulled up the good leg to stand on the neatly healed one to relax. She wouldn't eat on her own and had to be force-fed until weeks later she finally snatched a mouse tugged on a string through her building. With only one foot in working condition and a 70% mortality rate for juvenile raptors, Mika would have been a poor candidate to release to the wild. She has laid eggs each year and guards them for up to 60 days. Her name comes from that of the Formula 1 race car champion Mika Hakkinen, another hero.

the **Barred Owl**
male November 14, 1998

Graham the Barred Owl has been charming children and adults alike since November of 1998. He was found by a roadside near the airport in Missoula and probably bounced off a vehicle in a nighttime foray for mice. He was a young of the year at the time, judging from his plumage, and the collision left him blind in one eye. Young raptors have a very high mortality rate, and we decided that his chances of surviving in the wild were very poor. His first program was on Montana public radio. He was wonderful on his debut and continues to be to this day. Barred Owls have been moving into the Northwest only recently and are increasing in numbers. What a welcome new addition to our forests. They have a beautiful call or hoot that has been likened to the phase "Who cooks for you. Who cooks for you-ALL", (or in these politically correct days, "I'll cook today, You cook ToMORrow"). So when I asked my husband for a name regarding cooking he suggested "Graham", as in Kerr, the Galloping Gourmet. It's actually a very fitting title for this owl, as he loves his mice. We are very thrilled to have Graham on the teaching team.

the Golden Eagle
male January 25, 1999

Nigel the Golden Eagle joined us in January of 1999. He was retrieved in a field after having been shot in the right wing, and had a wing tip amputated as a result. He immediately proved to be spirited, and recovered his lost weight with a diet of ground squirrels and deer meat. We began his "manning" process with hand feedings and walks on the gloved forearm to perches in the field and yard. It is remarkable that a fierce wild raptor would adapt to captivity after such a traumatic experience, or put up with people at all. Nigel has a roommate, Max, and Nigel joins him on the lecture circuit, where they share up to 80 appearances a year. Nigel is fairly mellow and in programs he jumps from his travel box, is fed tidbits of meat off the elbow-length glove, then hops to a tall T-perch. Judging by his plumage and molt, Nigel was probably 5 or so years old when shot, just entering adulthood. With his calm demeanor and regal presence, he is a true beauty to behold.

the Northern Saw-whet Owl
male October 6, 1999

Buster the Northern Saw-whet Owl is a character indeed. In 1999 this youngster came to my sub-permittee rehabilitator as a near fledgling and candidate for release. After having him learn to fly and hunt in her basement, he was "hacked out" in her barn, or released with a daytime roost to return to. Sadly he was back after a few days with a wing droop. Buster lives in a building next to DotCom the Northern Pygmy Owl. He is the nocturnal hunter with the big eyes, compared to the smaller eyes of the daytime Pygmy. Buster is content with his two-mice-a-day diet, which is half his body weight, so he does "eat like a bird". Like most of our birds, he eats the heads off the mice first, like the frosting on the cupcake. The hindquarters he caches for later meals by draping them over limbs in his enclosure, which looks very decorative. He has provoked the hugest audience responses when he is lifted out of his box with cries of "he's so cute." We remind everyone that he is a predator and you wouldn't say that if you were, say, a mouse. We are very excited and fortunate to have this little fellow in the ranks.

the Northern Pygmy-Owl
male February 17, 2000

DotCom the Northern Pygmy-Owl is our our tiniest resident raptor. He is just over six inches long even with that relatively long tail, and he weighs as much as ten quarters (as compared to the nine-pound male Golden Eagles that are his neighbors). Dotcom is yet another victim of an automobile/bird interaction, in June of 2000, and has a wing injury. He enjoys several small mice each day that are placed on a shelf in the corner of his enclosure. In the wild he would catch birds up to twice his weight during twilight and daytime hunts. We often see Pygmy-Owls hanging around bird feeders, especially over the winter. These owls have markings on the back of their head that look like eyes to fool a potential attacker. Many people think that these small owls are just babies, so we inform them that when they leave the nest that is their full size. DotCom's name comes from the digital world that Raptors of the Rockies has joined, with our adventures in the computer world. We hope that this petite yet feisty owl will be content with his role as educator. He is by far the calmest Pygmy-Owl that we've ever encountered.

the American Kestrel
male June 20, 2000

JayDub is the latest Kestrel to join our team, and this little adult male is a beauty. He suffered a wing injury after a collision with a vehicle, like so many of the birds in our program. Roads are also the reason that so many are recovered, since people are constantly driving around and witness these encounters in the open, as compared to the problems that raptors run into where people are scarce. JayDub joined the program in June of 2000 and has been a welcome new member. He has limbs, posts, and a rope swing in his building, and he spends a lot of time in the sun. The name JayDub comes from the nickname of my high school art teacher in Cincinnati, Jack Walther, who, when I said I wanted to be an artist when I grew up, told me, "You ARE an artist." This diminutive falcon enjoys his feast of mice with chicken or quail meat. What a BIG personality for such a small bird.

the **Harris's Hawk**
female June 24, 2001

Deja the Harris's Hawk joined us as a fresh-out-of-the-chamber, captive-bred youngster and was placed on a falconry permit in 2001. She was sent by a breeder in upstate New York, and is the sister of Rio, who was killed the year before by a wild owl. We named her Deja, like Rio DEJA neiro, or better yet DEJA vu, with her strong similarities to her late sister. Harris's Hawks inhabit the very southern edges of the United States—Arizona, New Mexico, and Texas, with a range extending far in to central and South America. A dry country desert hawk, these birds have adapted to all climates and are prized for falconry around the world. Over the winter, Deja has a heated perch and a heating lamp to stand by, and we don't fly her at temperatures below 10 degrees farenheight. Otherwise she is flown daily, following along in the tree tops looking for pheasants below. She has been very instrumental in discussions about falconry. This species is bright and gregarious. They often hunt using teamwork between pairs and groups for greater success in securing prey. Deja is a personal favorite, usually.

the **Gyrfalcon**
male July 25, 2001

Gunnar the Gyrfalcon is a miracle bird, a young white "jerkin" that joined us in July of 2001. He was a gift from raptor breeder Bob Berry in Sheridan, Wyoming. That season had produced Gunnar, a gyrfalcon with developmental problems and impaired vision. He was saved from euthanasia because Bob hoped he would be comfortable with his disability as he got older, and this has held true. He was named Gunnar, a fine nordic name for a bird that is the national bird of Iceland. As an arctic native, he needs to have his building kept cool in the 100-degree heat of summer. First we installed shade cloth, then 2-inch R Board insulation in the rafters, 2 refelctive "space" blankets on the roof, a fan, then a water misting system, all keeping it a comfortable 85 degrees. Ironically he lives right next to the Harris's Hawk that loves the heat. Each day he joins us inside to attack and tear up balls of paper, or outside to bathe in the sprinkler or harass the Golden Retrievers. He has only tried to fly any distance once and crash landed, so has learned to hold his legs out to the sides for little flights. He is a playful and happy addition to our teaching team, and a truly remarkable bird.

Dulce

the **Peregrine Falcon**
female May 22, 2002

Dulce the Peregrine Falcon has just joined our family, and what a welcome addition. She came from Springhill Falcons of Bozeman in May of 2002. Our friend Skip Tubbs breeds Peregrines and we wanted a young bird to raise for falconry and education. Both of her parents are proven hunters with great field experience. After three eggs were layed by the female, they were pulled and incubated by Skip. These nestlings were hand reared for one week, then returned to the breeding chamber and parents for several weeks. At 23 days of age, she was pulled from the nest as a "late imprint," or one that wouldn't scream for food constantly. Dulce spent her first few weeks in our living room and then back porch and yard, finally losing her downy feathers at 38 days. I was with her every minute and training began when she was "hard penned" at 55 days when all of her feathers had grown in. Dulce is an *anatum* subspecies of Peregrine, or the continental kind that is native to the Rockies. Her name means "sweet" in Spanish, and that she is, sweet and smart and content, and we are so happy to have her here.

23 days old

27 days old

31 days old

36 days old

50 days old

Jeremy Puckett

Raptors of the Rockies

Raptor Topography

The term topography relates to mapping or charting, in this case the surface area of a raptor. Most of these terms relate to all birds. Notable exceptions include talon, the claw or nail on the raptor foot, and the hallux, or back toe, a piercing weapon. Also, the supercilliary ridge is unique; it helps protect the eyes, especially when hitting prey. The Falconiformes all have a "fierce" appearance due to this ridge. A falcon has a tomial notch on the beak and a bare patch of skin around the eye, or eye ring.

All of the feathers overlap like the shingles on a roof. Body or contour feathers are arranged in groups or tracts and distinct areas may be observed, like the scapulars on the shoulders and axillars on the underwing. Flight feathers on the wing each have a separate origin. The 10 primaries, or remiges, are attached to the manus or "hand," and the 12 or more secondaries are attached to the ulna. Rectrices refer to the tail feathers. Most of the owls are feathered all the way to the talons.

Glossary for the Raptors of the Rockies

adult – a full-grown individual with breeding potential and mature plumage

aerodynamic – streamlined and efficient in flight

allopreening – the process of one bird, using its bill to preen the feathers of another, usually between mated individuals

alula – group of feathers at the "thumb" that maintain a smooth flow of air across the upper wing surface

altricial – hatching bird with closed eyes, downy or naked, reliant on parent for heat and feeding

asychronous hatching – chicks hatching at different times, so that various sizes in nest with the oldest being the largest

asymmetrical – not exactly the same on either side from left to right

Aves – in biological taxonomy, the scientific class of all birds

axillar – the "armpit" region of a bird, and associated axillary feathers

avian – relating to birds

binocular vision – an overlap in sight between the left and right eye, allowing for greater depth perception

biology – the study of all living things

biomagnification – increasing concentrations of such things as toxins in the bodies of animals higher in a long food chain

bow trap – a round net trap set on the ground, half of which folds over on itself, often spring-loaded

brood – to provide heat and protection to nestlings; also the number of nestlings

cache – to store or hide uneaten portions of food to return to later

Cainism – one nest mate causing another to die, as in many Golden Eagles with two young (from the story of brothers Cain and Abel in the Bible)

camouflage – ways for an animal to remain disguised by blending in with their surroundings

carrion – the meat of animals found dead

casting – the indigestible parts of a meal, such as fur, feathers, and bones, that is coughed up in the form of a compact mass (called a pellet for owls)

"GHO"   Kate Phillips Davis

cere – the soft, bare patch of skin around the base of the bill in hawks, owls, parrots, and some pigeons

clutch – the total number of eggs laid by the female in one brood

cones – the photoreceptive cells in an animal's eye that distinguish sharp images and colors in day light

coniferous – evergreen trees like pine and fir that are cone-bearing, and keep their needles year-round (including larch trees, which lose their needles in the fall)

conspicuous – obvious and easy to see

convergent evolution – similar behaviors or physical traits in unrelated species that occur because of common selective pressures, like sharp talons in hawks and owls for hunting prey

cosmopolitan – species or groups that live all around the world

coverts – the small feathers that cover the longer wing and tail flight feathers

crepuscular – active at twilight (dawn and dusk)

"MEFE"   Kate Phillips Davis

facial disk – the ruff of feathers surrounding the eyes of many owl species that helps focus the sound and direct it to the ear openings; also found in harriers

Falconiformes – the order of birds that includes Ospreys, Old World vultures, hawks, eagles, falcons, and sometimes New World Vultures

falconry – the ancient sport of training raptors to hunt wild game in the field

falconry permit – the federal and state permission to keep a live raptor or raptors for the sport of falconry

fledge – to leave the nest as a young bird does; often still fed by parents

flyway – a concentration of migrating birds at certain favorable and predictable locations

fovea – the depression on the back surface of the eye where the image is focused and magnified

fratricide – siblings killing nest mates, especially in times of low prey availability

gape – the size of the open mouth of a bird

genetics – the science of how different plants and animals are related to each other, as in heredity

crop – a pouch in the esophagus at the throat of most birds that temporarily stores food before it moves into the first stomach; not found in owls

cryptic coloration – another term for camouflage, in which the plumage blends in with the surroundings to hide the bird when roosting and hunting

deciduous – broad-leaved trees; most species drop their leaves in autumn, like cottonwood and aspen

decurved – curving downwards at the tip, as in beak and talons

dihedral – in a shallow "v" shape, as in the soaring flight profile of many birds

diurnal – active during daylight hours

DNA – deoxyribonucleic acid, the genetic "building" blocks that determine inherited traits

DDT – an organophosphate chemical used to kill insect pests that harm agricultural crops; banned in the U.S in 1972

dorsal – regarding the back or upper side of an animal, where the spine is in vertebrates; opposite of ventral

ear tufts – the group of feathers on the top or sides of an owl's head, that form "horns"

eyrie – the cliff or ledge nest of raptors such as falcons and Golden Eagles (also called aerie)

"Saw-whet"   Kate Phillips Davis

Raptors of the Rockies

"Max"   Kate Phillips Davis

irruption – movement of a larger numbers of birds than normal for the winter, often associated with cycles of prey availability (as with hares and goshawks)

jerkin – the male of a Gyrfalcon (*Falco rusticolus*)

juvenile – a young animal before it is of breeding age; relating to traits of a juvenile, such as plumage

keel – the large pointed ridge of bone on the sternum of birds for flight muscle attachment; the breast bone

kleptoparasitism – stealing food from another animal; also called pirating

low wing-loading – having a greater wing and tail surface area compared to a lower body weight, as in the buteo hawks and vultures, best for soaring flight and catching thermals

malar – the area on the face of a bird just below the eye, as in the malar stripe of a falcon

mammalogy – the scientific study of mammals

manning – the process of taming a bird to be comfortable around the handler or other people

gizzard – the second of two stomachs of a bird, where the pellet or casting is formed, the ventriculous

hallux – the back toe of a bird

high wing-loading – having a lower wing and tail surface area compared to a higher body weight; better for high speed flight as in falcons

hooked beak – upper mandible curves down below lower; primarily for cutting meat and separating bony parts

hybridization – the interbreeding of two different animal species, usually quite rare in nature; a hybrid is the product of this

ichthyology – the scientific study of fishes

immature – young bird no longer in care of its parents and often in different plumage from adult; some can breed while still in immature plumage; different from a juvenile that is still too young to breed

imprinting – identification of a juvenile bird with its parents; human interference with a very young bird may cause it to always associate with people and not other birds

"Peregrine Falcon"   Kate Phillips Davis

Glossary for the Raptors of the Rockies

mantle – to cover food with spread wings and tail, so as to hide it from others that may want to steal it

manus – the "hand" region on the wing, on which the primary feathers are attached

metabolism – the chemical processes of an animal, such as digestion and respiration

migration – the movement of animals from one area to another when the seasons change

mobbing – the defensive behavior of prey birds around predators, often in flocks, noisily indicating that danger is present

molt – the process by which old feathers are replaced, largely during the summer months

morph – a type of color pattern within a species

morphology - the study of structure and anatomy in animals

motion parallax- the fact that nearby objects appear to move more quickly relative to the viewer than those farther away, allowing depth perception; birds take advantage of this by bobbing and swiveling their heads to judge distance when hunting

Nearctic – the northern portion of the Western Hemisphere

New World – the Western Hemisphere of North and South America

niche – the individual place an animal may live in the environment

nictitating membrane – the third eyelid of all birds, keeping the eye moist and protecting the surface

nocturnal – active at night

Old World – the Eastern Hemisphere of Europe, Asia, Africa, Australia

ornithology – the scientific study of birds

osteology – the scientific study of bones and their structure

oviparous – animals that lay eggs, like birds and reptiles

patagial – the area on the wing with the patagium, or sheet of skin that connects the shoulder area to the manus

pellet - compact ball containing the bones and fur or feathers of an owl's meal, which is regurgitated, or coughed up

plumage – the feathers of a bird, often indicating age and sex

polyandry – a female mating with more than one male during breeding season, rare in the wild

polygamy – having more than one mate in breeding season

"Flipper"   Kate Phillips Davis

polygyny – a male mating with more than one female during the breeding season; sometimes occurring in harriers and some owl species

predatory – relating to a lifestyle of killing other animals for food

preening – straightening and cleaning the feathers by drawing the feather length through the beak, and adding oil from the uropygial or oil gland at the base of the tail to waterproof the plumage

quarry – the intended prey of a hunter

race – another word for subspecies

range – the geographic region where an animal is found at a certain time of year, as in winter range, breeding range, year-round range

raptor – a bird of prey, including the orders of hawks and owls, from the Latin *raptare* meaning "to seize and carry away"

rectrices – the tail feathers, usually numbering twelve in raptors

remiges – the flight feathers on the wing; primaries and secondaries

Glossary for the Raptors of the Rockies

rods – the photoreceptive cells in an animal's eye that best work in low light, and are more numerous in owls

roost – the time a bird spends inactive, or sleeping; and the place a bird spends this time

search image – a fixed idea in the bird's brain of what prey items look like in the field, like a pheasant in the grass, often from a successful hunting experience

scapulars – the group of nonflight feathers that cover the shoulders

sclera – the bones radiating from the skull around the eyes, causing the fixed stare of owls

sexual dimorphism – size difference between the sexes, with reverse sexual dimorphism in most raptors in which the female is larger than the male

shelterbelt – a natural or planted line of trees or shrubs growing in open country and serving as a windbreak

siblicide – the killing of a nest mate, usually to allow for more food to the survivor

spicules – spikey protrusions on the feet to better help in gripping prey, as found in Ospreys for holding slippery fish

still-hunt – to hunt from a perch

Strigiformes – the order of birds that comprises owls

subadult – the young of a bird that takes more than one year to reach breeding age

subspecies – a geographic variation of a species marked by a third scientific name, like *Buteo jamaicensis calurus* or the Red-tailed Hawk of the West

supercilliary – the area above the eye

talon – the curved "claws" on the toes of a bird of prey

taiga – the moist spruce/fir habitat between the tundra to the north and coniferous forest to the south

tarsus (plural tarsi) – the tarsometatarsus, or long foot bone, from which the toes originate

tiercel – the male of birds in the falcon family, so called because they are one-third smaller than the females

"Prairie Falcon and Horned Larks"   Kate Phillips Davis

tomial notch – a notch on the upper beak of falcons to help break the neck of the prey and separate bones at joints

taxonomy – a way to classify living organisms according to who is related to whom

thermal – the process of ground air rising due to lighter weight when heated, which creates soaring conditions for birds

triangulate – the method of locating prey by owls using hearing due to asymmetrical ear openings

tundra – the treeless, frozen ground of the arctic regions

uropygial gland – the oil gland at the base of the tail used in preening

ventral – relating to the underside or abdominal side of animals, as opposed to dorsal

wing-loading – the ratio of a bird's weight, or body mass, to wing surface area.

zoology – the scientific study of all animals, from invertebrates to mammals

"Bald Eagle"   Kate Phillips Davis

Raptor Resources

The Peregrine Fund
5668 West Flying Hawk Lane
Boise, ID 83709
www.peregrinefund.org

The Raptor Center
at the University of Minnesota
Gappert Raptor Building
1920 Fitch Ave.
St. Paul, MN 55108
www.raptor.cvm.umn.edu

Cornell Lab of Ornithology
159 Sapsucker Woods Road
Ithaca, NY 14850
www.ornith.cornell.edu

Hawk Migration Association
of North America (HMANA)
P.O. Box 822
Boonton, NJ 07005-0822
www.hmana.org

HawkWatch International
1800 S. West Temple, Suite 226
Salt Lake City, UT 84115
www.hawkwatch.org

Hawk Mountain Sanctuary
1700 Hawk Mountain Rd.
Kempton, PA 19529-9449
www.hawkmountain.org

For ornithology books:

Buteo Books
3130 Laurel Rd.
Shipman, VA 22971
www.buteobooks.com

"Crackity Jones"   Kate Phillips Davis

References

Brown, Leslie and Amadon, Dean. *Eagles, Hawks & Falcons of the World*. Secaucus, N.J: The Wellfleet Press, 1989.

Clark, William S., and Brian K. Wheeler. *Peterson Field Guide: Hawks*. Boston: Houghton Mifflin Company, 1987.

Collard, Sneed B. *Birds of Prey*. New York: Grolier Publishing, 1999.

Cox, Randall T. *Birder's Dictionary*. Helena Montana: Falcon Press, 1996.

Dunn, Jon L. *National Geographic Field Guide to the Birds of North America: Revised and Updated*. 3d ed. Washington, D.C.: National Geographic Society, 1999.

Dunne, Pete, David Sibley, and Clay Sutton. *Hawks in Flight*. Boston: Houghton Mifflin Company, 1988.

Ehrlich, Paul, David Dobkin, and Darryl Wheye. *The Birder's Handbook*. New York: Simon & Schuster Inc., 1988.

Grossman, Mary Louise, and John Hamlet, *Birds of Prey of the World*, New York:Clarkson N. Potter, Inc. 1964.

Hoyo, Josep del, et al., *Handbook of the Birds of the World: Barn Owls to Hummingbirds*. Vol. 5. Barcelona: Lynx Edicions, 1999.

Josep del, et al., *Handbook of the Birds of the World: New World Vultures to Guineafowl*. Vol. 2. Barcelona: Lynx Edicions, 1994.

Hume, Bob, and Trevor Boyer. *Owls of the World*. London: Collins and Brown Ltd., 1991.

Johnsgard, Paul A. *Hawks, Eagles & Falcons of North America*. Washington, D.C.: The Smithsonian Institution, 1990.

Johnsgard, Paul A. *North American Owls*. Washington, D.C.: The Smithsonian Institution, 1988.

Kaufman, Kenn. *Lives of North American Birds*. Boston: Houghton Mifflin Company, 1996.

Lloyd, Glenys, and Derek Lloyd. *Bantam Knowledge Through Color: Birds of Prey*. New York: Grosset and Dunlap, 1971.

Newton, Ian. *Population Ecology of Raptors*. Hertfordshire, UK: T&A D Poyser Ltd, 1979.

Robbins, Chandler S., et al. *Birds of North America: A Guide to Field Identification*. Wisconson: Golden Book Publishing Co., 1983.

Sibley, David Allen. *The Sibley Guide to Birds*. New York: Alfred A. Knopf: 2000.

Sibley, David Allen. *The Sibley Guide to Bird Life and Behavior*. New York: Alfred A. Knopf, 2001.

Weidensaul, Scott. *Raptors: The Birds of Prey*. Hong Kong: Lyons & Buford, 1996.

Wheeler, Brian K., and William S. Clark. *A Photographic Guide to North American Raptors*. San Diego: Academic Press, 1996.

"Winter Pygmy" Kate Phillips Davis

Index

Page references in **boldface italics** refer to photos or illustrations.

"A Clockwork Owl"   Kate Phillips Davis

Index continued

Ron Austing

Kate Davis

Executive Director

We often hear people talk about their life-long commitment to their profession. The phrase most certainly fits for educator, author, and artist Kate Davis, who began at an early age. By 1973 her fascination with wildlife was firmly ensconced when she joined the Jr. Zoologist Club at the Cincinnati Zoo. There she began to take care of raccoons, red and gray foxes, and her first birds of prey. She began to conduct her first education programs at the age of 14 at the zoo and for area schools and camps. What began as a childhood fascination has grown into a full-time, all-consuming, live-eat-breathe-never-leave-town-because-I've-got-to-take-care-of-the-birds passion. By training she is a zoologist, having earned her degree with honors from the University of Montana in 1982. By choice she is an educator, with a commitment that is obviously enduring. And finally, by mistake, she is an administrator, having formalized her work by creating the nonprofit organization Raptors of the Rockies, for which she now serves as Executive Director and Chief Raptor—Master of the Flock. Resident raptors are the source for Kate's depictions in drawings, etchings, monotype prints, paintings, and sculptures.

She is an internationally recognized artist, having been selected for the Woodson Birds in Art Exhibition. Kate and birds appeared on national television with the PBS show *BirdWatch*. She was named Environmental Educator of the Year 2000 by Montana Audubon and Conservation Educator of the Year 2002 by the Montana Wildlife Federation. Program numbers now exceed 700 for more than 60,000 audience participants since 1988, and more than 80 schools have enjoyed a visit from the teaching team of permanently disabled birds of prey. With the Raptors' recent move to permanent ground in Montana's Bitterroot Valley, their future is very bright indeed.

The word *raptor* comes from the Latin word for seize & carry away. Kate is clearly a raptor, for she has seized a passion for wildlife and has carried it away into a life of creative educational service to birds.

Robert Petty, Director of Centers and Education
Montana Audubon

Help Support the Raptors of the Rockies Educational Program

Raptors of the Rockies continues to educate schools and the public through live appearances, our web site, and quarterly newsletters. This nonprofit is primarily funded by program fees, adoptions, grants, sales of merchandise, and donations.

We could use your help to ensure a bright future.

Costs in caring for these raptors include :

Food Purchases: whole chickens, quails, ground squirrels, rabbits, heart meat, and mice, along with donated game meat

Freezer Space: running nonstop

Enclosure Maintenance: roofing, perches, shade cloth, and sand, gravel, pine needle floors, fencing, heat lamps, and fans

Daily Care: 365 days a year—food procurement, butchering , feeding, watering, keeping things tidy and healthy

Insurance: to allow birds in the classroom

Driving to Programs: over 5,000 miles a year

Paperwork: to maintain four federal permits, program information, budget, correspondence, special events, sponsorship solicitation

GRAHAM
THE BARRED
OWL

Please know that your sponsorship helps make this educational service possible.
All contributions are tax deductible (501(c)(3) non-profit) and your support is very much appreciated!

Kate Davis, Executive Director

Raptors of the Rockies

P.O.Box 250, Florence, Montana 59833
raptors@montana.com www.raptorsoftherockies.org

✂ ···

YES, I want to sponsor the Raptors of the Rockies with this tax-deductible contribution:

Name _____

Address _____

Contribution amount _____

or   Additional copies of Raptors of the Rockies at $16.00 each plus $4 shipping and handling

Number of copies _____ TOTAL $_____

For your donation you will receive a receipt for your tax records, and join the mailing list for our quarterly newletter **Raptor Round-Up** for one year.

1341

We THANK YOU, and Welcome to the Family!